OPTIMIZATION AND CONTROL OF ELECTRICAL MACHINES

Edited by **Abdel Ghani Aissaoui, Ahmed Tahour** and **Ilhami Colak**

Optimization and Control of Electrical Machines

http://dx.doi.org/10.5772/intechopen.71523

Edited by Abdel Ghani Aissaoui, Ahmed Tahour and Ilhami Colak

Contributors

Jorge De Almeida Brito Júnior, Manoel Nascimento, Jandecy Cabral, Jorge Moya, Marcus Vinicius Alves Nunes, Carlos Alberto Oliveira De Freitas, Milton Fonseca Júnior, Edson Farias De Oliveira, David Barbosa Alencar, Nadime Mustafa Moraes, Haroldo Melo De Oliveira, Tirso Lorenzo Reyes Carvajal, Dr. Venu Madhav Gopala, Obulesu Y. P., Qiang Gao, Greg Asher, Mark Sumner, Mandar Padmakar Bhawalkar, N. Gopalakrishnan, Yogesh Nerkara, Abdel Ghani Aissaoui

Notice

Statements and opinions expressed in the chapters are these of the individual contributors and not necessarily those of the editors or publisher. No responsibility is accepted for the accuracy of information contained in the published chapters. The publisher assumes no responsibility for any damage or injury to persons or property arising out of the use of any materials, instructions, methods or ideas contained in the book.

First published in London, United Kingdom, 2018 by IntechOpen

IntechOpen is the global imprint of INTECHOPEN LIMITED, registered in England and Wales, registration number: 11086078, The Shard, 25th floor, 32 London Bridge Street

London, SE19SG – United Kingdom

Printed in Croatia

British Library Cataloguing-in-Publication Data

A catalogue record for this book is available from the British Library

Additional hard copies can be obtained from orders@intechopen.com

Optimization and Control of Electrical Machines, Edited by Abdel Ghani Aissaoui, Ahmed Tahour and Ilhami Colak

p. cm.

Print ISBN 978-1-78923-452-7

Online ISBN 978-1-78923-453-4

We are IntechOpen,
the world's leading publisher of
Open Access books
Built by scientists, for scientists

3,600+
Open access books available

113,000+
International authors and editors

115M+
Downloads

Our authors are among the

151
Countries delivered to

Top 1%
most cited scientists

12.2%
Contributors from top 500 universities

CLARIVATE ANALYTICS
BOOK
CITATION
INDEX
INDEXED

WEB OF SCIENCE™

Selection of our books indexed in the Book Citation Index
in Web of Science™ Core Collection (BKCI)

Interested in publishing with us?
Contact book.department@intechopen.com

Numbers displayed above are based on latest data collected.
For more information visit www.intechopen.com

Meet the editors

Abdel Ghani Aissaoui is a full Professor of Electrical Engineering at the University of Bechar (Algeria). He was born in 1969 in Naama, Algeria. He received his BS degree in 1993, MS degree in 1997, PhD degree in 2007, all from the Electrical Engineering Institute of Djillali Liabes, University of Sidi Bel Abbes (Algeria). He is an active member of IRECOM (Interaction Réseaux Electriques - COnvertisseurs Machines) Laboratory and an IEEE senior member. He has edited many international journals (IJET, RSE, MER, IJECE, etc.) and serves as a reviewer in international journals (IJAC, ECPS, COMPEL, etc.). He serves as a member in a technical committee (TPC) and reviewer in international conferences (CHUSER 2011, SHUSER 2012, PECON 2012, SAI 2013, SCSE 2013, SDM 2014, SEB 2014, PEMC 2014, PEAM 2014, SEB (2014, 2015), ICRERA 2015, etc.). His current research interests include power electronics, control of electrical machines, artificial intelligence and renewable energies.

Ahmed Tahour was born in 1972 in Ouled Mimoun, Tlemcen, Algeria. He received his BS degree in electrical engineering in 1996, MS degree in 1999 and PhD degree in 2007, all from the Electrical Engineering Institute of the University of Sidi Bel Abbes (Algeria). He is currently a Professor of Electrical Engineering at the University of Mascara (Algeria). His current research interests include power electronics, control of electrical machines and renewable energies.

Ilhami Colak was born in 1962 in Turkey. He received his diploma in electrical engineering from Gazi University, Turkey, in 1985. Then, he did his MSc degree in electrical engineering in the field of *Speed Control of Wound Rotor Induction Machines Using Semiconductor Devices* at Gazi University in 1991. After that, he received his MPhil degree from Birmingham University in England by preparing a thesis on *High Frequency Resonant DC Link Inverters* in 1991. He received his PhD degree from Aston University in England on *Mixed Frequency Testing of Induction Machines Using Inverters* in 1994. He became an Assistant Professor, an Associate Professor, and a full Professor in 1995, 1999 and 2005, respectively.

He has published more than 225 papers on different subjects, including electrical machines, drive systems, machine learning, reactive power compensation, inverters, converters, artificial neural networks, distance learning, automation, renewable energy sources and smart grids. More than 86 of his papers have been cited in the SCI database of Thomson Reuters. His papers have received more than 445 citations. He has supervised 19 MSc students and 13 PhD students. He is a member of IEEE, IES, IAS, PELS and PES. He is also a member of the PEMC Council. He has organized 54 international conferences and workshops. In the last ten years, he has concentrated his studies on renewable energy and smart grids by publishing papers, journals (www.ijrer.org), and organizing international IEEE sponsored conferences (www.icrera.org). He is also the editor-in-chief of International Engineering Technologies (http://dergipark.ulakbim.gov.tr/ijet) and one of the editors of the Journal of Power Electronics (http://www.jpels.org). He has 1 international and 3 national patents. He also spent around 3 years at the European Commission Research Centre (JRC) as an expert in the field of smart grids in the Netherlands. He is currently holding the positions of vice-rector and dean of the Engineering and Architecture Faculty of Nisantasi University.

Contents

Preface

Electrical drive systems were developed in the last century. The majority of all drive systems are electrical and this tendency is increasing. The main parts of these systems are electrical machines. With the advent of power electronics, there are new possibilities for the control of electrical machines with variable speed. The technical performance and economical design of electrical machines have opened a new philosophy of drive applications. Electrical machines have undergone a huge development. New concepts in design and control allow for the expansion of their applications in different fields. Electrical machines are considered important components in many industrial applications such as power systems, manufacturing plants, power plants, electrical vehicles and home appliances.

These machines are usually manufactured through mass-production techniques. Their performance can be affected by manufacturing processes and different operational conditions (e.g., temperature). As a requirement of quality control, an optimal design process is often applied to minimize the influence of uncertainties on the machine's performance. As a consequence, an optimal design system can be implemented for the analysis and design of electrical machines to minimize the effects of manufacturing errors in both iron and permanent magnets.

Control is very important; it allows having a good functioning of the system. Different strategies of control can be applied to electrical machines depending on the necessary goals. The use of advanced control techniques and new technology brings the system into its optimal operating mode.

The goal of this book is to present recent works on design and control related to electrical machines. The developments of this field happened quickly, accompanied with many difficulties that require solutions. Different solutions are proposed based on new techniques of control, design and advanced technology products.

This book is divided into two parts. The first part is devoted to the design of different optimization techniques for electrical systems:

- The introductory chapter describes electrical machines.
- In the first main chapter of Part 1, several optimization techniques (e.g., Simulated Annealing, Ant-Lion, Dragonfly, NSGAII, and Differential Evolution) are analyzed and discussed to solve the economic-emission load dispatch problem. A comparison of all approaches and their results is offered through a case study.
- The second chapter highlights various optimization methods and discusses the suitability of non-linear programming methods and recent stochastic or population-based methods for optimal design of brushless doubly fed reluctance machines (BDFRM).

The second part is dedicated to the control of induction machines:

- The first chapter presents the zero and low speed sensorless control of induction machines using only fundamental pulse width modulation (PWM) waveform excitation. A position sensorless method is used, which only relies on the fundamental PWM waveforms to excite saliency. This method is essentially based on saliency detection, and therefore derivation of the rotor position is possible at low and zero speeds.
- In the second chapter, the development of a rotor flux reference generation control strategy is explained. This control strategy is used with the DTC scheme to eliminate the high peaks in torque with reduced stator and rotor currents and also to eliminate the necessity of a crowbar during low voltage dip. The scheme allows for the DFIG to be connected to the grid even during a fault.

The chapters of this book present recent works in the design, control, and applications of electrical machines and their importance. The aim of this book is to present the new trends of research on electrical machines.

We hope that the readers will find this book to be a significant source of knowledge and reference for the future years.

Prof. Dr. AbdelGhani Aissaoui
Electrical Department
Faculty of Technology
University Tahri Mohamed of Bechar (UTMB)
Bechar, Algeria

Prof. Dr. Ahmed Tahour
Electrical Department
Faculty of Technology
University of Mascara
Mascara, Algeria

Prof. Dr. Ilhami Colak
Engineering and Architecture Faculty
Nisantasi University
Istanbul, Turkey

Optimization Techniques in Electrical Systems

Introductory Chapter: Introduction to the Design and the Control of Electrical Machines

Abdel Ghani Aissaoui

Additional information is available at the end of the chapter

http://dx.doi.org/10.5772/intechopen.78772

1. Introduction

In the last century, electrical machines have been the subject of a huge development. New concepts in design and control allow expanding their applications in different fields. They are considered important components in many industrial applications as: power systems, manufactories, power plants, electrical vehicles, and home appliances.

There are several types of electrical machines; we can find synchronous machines, induction machines, direct current (DC) machines, reluctance synchronous machines, transformers, etc. (**Figure 1**).

The electrical machines are incorporated into the process of energy conversion in the generation, transmission, and consumption of electric power. In a power station, turbine generator converts the energy coming from the combustion of coal, natural gas, etc. into electric energy

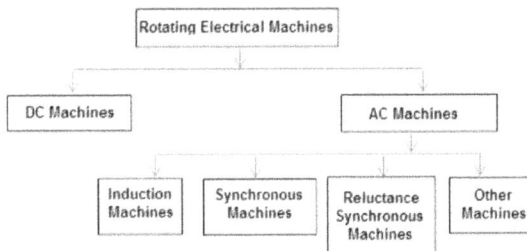

Figure 1. Different types of electrical machines.

that is transmitted to consumers: motors whose mechanical energy drive machines in industry, homes, traffics, etc.

Most applications are interested in rotating electrical machines. The rotating electrical machine can operate, without constructional changes, as a motor or generator, since the energy direction of an electrical machine is reversible (**Figure 2**).

Electrical machines can be classified according to the torque producing mechanism and their magnetic interactions. The first class based on the torque producing mechanism machines is classified into two types, one is alignment torque producing machines such as DC machines, induction, and synchronous machines and the second is the reluctance torque producing machine, for example, switched reluctance machines. The second class based on the magnetic interactions machines is classified as inductive-interactive type machines, for example, induction machines, synchronous machines, and DC machines, and variable reluctance type machines, for example, switched reluctance machines [1].

The electrical drive systems were developed based on the use of electrical machines. The majority of all drive systems are electrical drives with growing tendency. Electrical drive systems do not have a power density as high as pneumatic or hydraulic systems. Electrical motors are bulky and heavy in comparison to these competitors.

Electrical drives are considered for three reasons superior to other drive systems, such as pneumatic and hydraulic systems:

- Cleanliness of the energy supply

- Dynamics of control

- High efficiency of electromechanical power conversion

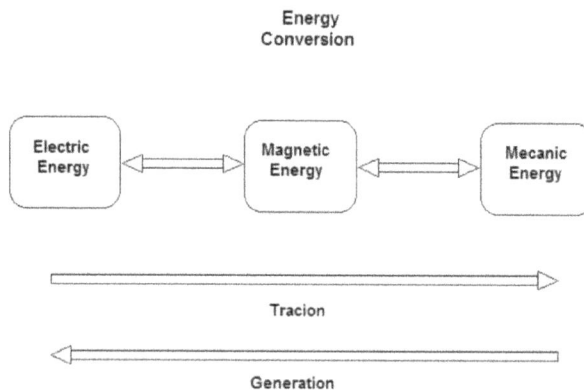

Figure 2. Conversion energy in electrical machines.

The main part of these systems is electrical rotating machine. With the advent of power electronics, new possibilities appear for electrical machines with variable speed. Their technical performance and economical design opened a new philosophy of drive applications.

The control of electrical power today is possible within short time for megawatts. It can be controlled so fast than any other form of energy.

The energy conversion between electrical and mechanical power is performed by the electrical machine in both directions.

Electrical machines can be used for different ranges of speed. It can be used as motor particularly in traction, electrical vehicles, etc. or as generators in power station, wind turbines, etc.

2. Design

The electrical machines are usually manufactured through mass-production techniques; their performances can be affected by manufacturing processes and different operational conditions (e.g., temperature). As a requirement of quality control, a robust design process is often applied to minimize the influence of uncertainties on the machine performance.

However, the conventional computer simulation cannot reflect the influences of the environmental uncertainties directly. The input data of the numerical model are usually the geometries of the modeled device, the material properties, and the uncertainties in both must be taken into account. While most of the works in the robust design of electromagnetic devices focus on the uncertainties in the geometries [2–4], only a few efforts have been conducted on the influences of the material uncertainties. In electromagnetic field computing, the nonlinear behavior of the constitutive laws of ferromagnetic materials is usually obtained by B-H curves. For ferromagnetic materials, [5] constructed a stochastic material model using the uncertainties of the measured points to characterize a nonlinear B-H curve. In [6], a global sensitivity analysis was applied to study the variance of the predicted behavior of a turbo-alternator with respect to material uncertainties.

All the above stochastic material models have formed a solid foundation for the study of material uncertainties in the electromagnetic design. A robust design system can then be implemented for the analysis and design of electrical machines in order to minimize the effects of manufacturing errors in both iron and permanent magnets [7].

There are some general design methods, which can be applied in terms of different disciplines/domains: electromagnetic design, thermal design, structural design, multi-physics design, material design, and manufacturing process design.

Electromagnetic design: The principle of operation of electrical machines is based on the electromagnetic theory. Electromagnetic design is based on the calculation of magnetic field and its distribution in the electrical machines, which allowed to compute some basic electromagnetic parameters including winding inductance and the evaluation of some performances, such as electromagnetic force, power loss, and efficiency. To obtain the magnetic field, there are three main kinds of analysis methods, analytical method, magnetic circuit method, and finite element method (FEM) [8–10, 14, 15]. Meanwhile, power losses and efficiency are two important performance indexes for electrical machines.

Thermal design and structural design: these methods can be applied after the accomplishment of the electromagnetic design. The thermal design method aim is to compute the temperature distribution in the machine based on the heat obtained from the electromagnetic analysis. There are popular methods for the thermal analysis of electrical machines. They are computational fluid dynamics (CFD), FEM, and nodal method. Structural design aims to consider the stress and deformation of the machine under the electromagnetic and thermal analyses. Structural design can be conducted based on FEM as well [12, 13].

Multi-physics design: It aims to calculate the electromagnetic characteristics, temperature distribution, structural stress, vibration noise, and coupled performances of electrical machines based on a uniform model [8, 9, 10, 11]. The FEM has been widely employed as a powerful tool for the multi-physics design and analysis of electrical machines. It can be used to analyze the coupled field in machines, such as electromagnetic structure and thermal structure.

Material design: The type of material is important for the electromagnetic, thermal, and structural designs of electrical machines. Nowadays, new developed magnetic materials like soft magnetic composite (SMC), amorphous and grain-oriented silicon steel show better characteristics, such as high saturation flux density, low specific losses, and low manufacturing cost. They can be employed to design motors with new topologies, higher efficiency, and/or low manufacturing cost [16, 17].

Manufacturing process design: Manufacturing method design is also important in the design stage of electrical machines, which will influence their manufacturing quality and actual performances in operation. To obtain the best performances, some designs have complex structures which can be difficult for manufacturing.

With a good knowledge of the magnetic characteristics and manufacturing methods, we can fully exploit all the performances of the designed motors. A good motor design should be done in terms of both output performances and manufacturing abilities [18].

3. Optimization

Optimization is a very popular term in modern design of electrical machines and devices due to the competition in the world markets, increased cost of electrical energy, and pressures for its conservation.

Optimization helps designers to push the existing invisible design boundaries while using available materials and technology. The objective of the optimization process is usually to minimize either the initial cost of the machine or its lifetime cost including the cost of lost energy. Other objectives such as mass minimization or efficiency maximization may also be appropriate in some situations [19, 20].

This can be explained and understood through the words of Miller [21]: "To a WISE engineer, optimal design means a compromise between conflicting factors, often producing an imperfect result from optimistic aspirations."

Most of the metaheuristic techniques can be used to solve global optimization problems with nonlinear constraint by using metaheuristic algorithms; there is a high possibility to determine a near optimal solution, which can be considered by designer and engineering as a global optimum [22].

One of the most promising algorithms from the class of evolutionary algorithms widely used in the field of electric machines is Differential Evolution (DE) [22–24] first introduced by Price and Storn [25] in 1995. The algorithm was later improved and named Generalized Differential Evolution (GDE) (extended DE for constrained multi-objective optimization) by Lampinen and Zelinka [25, 26].

Variety of other algorithms is used in electric machine design optimization: Genetic Algorithm (GA) [27, 28], Particle Swarm Optimization (PSO) [29, 30], Simulated Annealing (SA) [31], etc. Authors in [32] compared GA, SA, and DE on the design optimization of permanent magnet motor and authors in [32] compared DE, GA, and PSO on the design optimization of microstrip antennas. Both groups agree that the DE performance is the best. In [33, 34], PSO and GA were compared and PSO was found computationally more effective with slightly better objective function value reached. In [35], it is shown how PSO performs better than GA so some authors decided to use hybrid GA-PSO method [36, 37].

4. Control of electrical machines

The control of electrical machines has been the subject of great progress, due to the development and advancement in the field of power electronic devices, digital signal processing, the informatics tools, and advanced control techniques.

The energy conversion between electrical and mechanical power is performed by the electrical machine in both directions. The control of this energy is very important. In the case of motors, we control the electric power consumed, and in the case of generator, we control the electric power generated. The control of electrical machines can be expanded to other variables as speed, voltages, currents, flux, torque, etc. **Figure 3** shows the control structure of electrical machines.

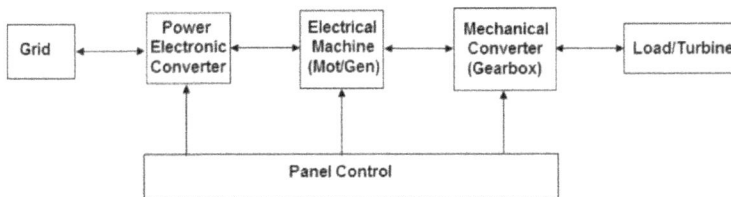

Figure 3. Control structure of electrical machines.

The electrical drive systems are based mostly on electrical machines. These machines can be designed to operate at different speeds: high, medium, and low speed. According to these benefits, the use of electrical machines with variable speed is very important in the field of power station, wind turbine, electrical vehicles, etc.

The electrical machines with high speed is in continuous evolution for a number of applications, including aero engine spools, electrical turbo-compounding systems, electrical spindles for milling cutters and grinding, helicopter and racing engines, turbochargers, fuel pumps, etc. These applications have typical operational speeds of over 10,000 r/min.

In the control design, we follow the next steps:

- Modeling: The plant can be described in the form of some mathematical equations. These equations constitute the mathematical model of the plant. A plant model should produce the same output response as the plant for the same inputs.

- Controller design: The controller is designed to meet the performance requirements for the plant model.

- Implementation: The implementation can be done using a digital computer. Its efficiency depends on the type of computer available, the type of interface devices between the computer and the plant, software tools, etc.

Figure 4 shows different steps of control design.

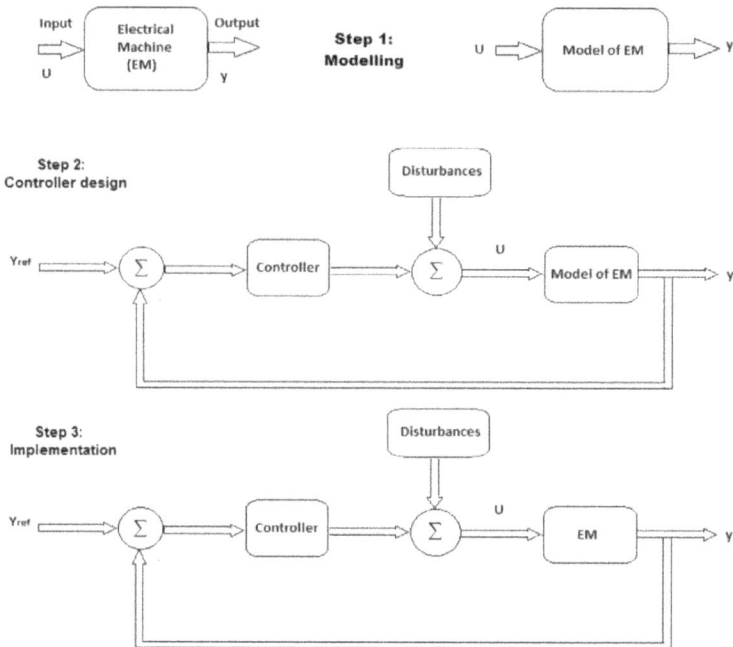

Figure 4. The steps of control design.

The control of these machines is very complicated. We need to represent them by mathematical models. Their models are defined by coupled and nonlinear equation systems.

Some mathematical transformations can be used to simplify the form of these models as Park transformation, Clark transformation, etc. The analytic approach of these equations is very complicated. To solve these systems of equations, the numerical methods are recommended.

Following the obtained models many strategies of controls were developed such as vector control, direct torque control, direct power control, etc., and these strategies aim to give more flexibility to the control systems.

New techniques of control were developed based on the powerful control theory and artificial intelligence tools. Many control techniques were studied and applied to electrical machine, which we can find variable structure control, model reference adaptive control, adaptive pole placement control, predictive control, backstepping control, etc. The use of artificial intelligence techniques has been the subject of many recent researches. The most famous are the fuzzy logic control, the neuronal control, the neuro fuzzy control, etc. In the literature, we can find many researches on these topics.

In [38], a model reference adaptive control-based estimated algorithm was proposed for online multi-parameter identification. In [39], MRAS observer was designed for the field oriented control of DFIG. Authors in [40] give an overview of model predictive control for induction motor drives. In [41], cascaded nonlinear predictive control was proposed for the control of induction motor. Backstepping controller was proposed in [42] for induction machine. An adaptive backstepping sliding mode controller was presented in [43]. A fuzzy logic controller was developed for a switched reluctance motor in [44].

Author details

Abdel Ghani Aissaoui

Address all correspondence to: irecom_aissaoui@yahoo.fr

Electrical Department, Faculty of Technology, University of Tahri Mohamed of Bechar, Algeria

References

[1] Rajendran N. A survey on electrical machines for variable speed applications. Indian Journal of Science and Technology. 2015;**8**(31). DOI: 10.17485/ijst/2015/v8i31/84306

[2] Yoon S-B, Jung I-S, Hyun D-S, Hong J-P, Kim Y-J. Robust shape optimization of electromechanical devices. IEEE Transactions on Magnetics. 1999;**35**:1710-1713

[3] Alotto P, Magele C, Renhart W, Weber A, Steiner G. Robust target functions in electromagnetic design. COMPEL: The International Journal for Computation and Mathematics in Electrical and Electronic Engineering. 2003;**22**:549-560

[4]   Steiner G, Weber A, Magele C. Managing uncertainties in electromagnetic design prob-
      lems with robust optimization. IEEE Transactions on Magnetics. 2004;**40**:1094-1099

[5]   Bartel A, de Gersem H, Hulsmann T, Romer U, Schops S, Weiland T. Quantification of
      uncertainty in the field quality of magnets originating from material measurements.
      IEEE Transactions on Magnetics. 2013;**49**:2367-2370

[6]   Mac D, Clenet S, Beddek K, Chevallier L, Korecki J, Moreau O, Thomas P. Influence of
      uncertainties on the B (H) curves on the flux linkage of a turboalternator. International
      Journal of Numerical Modelling: Electronic Networks, Devices and Fields. 2014;**27**:385-399

[7]   Li M, Mohammadi MH, Rahman T, Lowther D. Analysis and design of electrical
      machines with material uncertainties in iron and permanent magnet. COMPEL–The
      International Journal for Computation and Mathematics in Electrical and Electronic
      Engineering. 2017;**36**(5):1326-1337

[8]   Lei G, Zhu JG, Guo YG. Multidisciplinary Design Optimization Methods for Electrical
      Machines and Drive Systems. Berlin/Heidelberg, Germany: Springer-Verlag; 2016. ISBN
      978-3-662-49269-7

[9]   Lei G, Liu CC, Guo YG, Zhu JG. Multidisciplinary design analysis and optimization for
      a PM transverse flux machine with soft magnetic composite core. IEEE Transactions on
      Magnetics. 2015;**51**(11)

[10] Lei G, Liu CC, Zhu JG, Guo YG. Robust multidisciplinary design optimization of PM
      machines with soft magnetic composite cores for batch production. IEEE Transactions
      on Magnetics. 2016;**52**(3)

[11] Kreuawan S, Gillon F, Brochet P. Optimal design of permanent magnet motor using mul-
      tidisciplinary design optimization. In: Proceedings of the 18th International Conference
      on Electrical Machines; Pattaya, Thailand; 25-28 October, 2015. pp. 1-6

[12] Lin F, Zuo S, Wu X. Electromagnetic vibration and noise analysis of permanent magnet
      synchronous motor with different slot-pole combinations. IET Electric Power Appli-
      cations. 2016;**10**:900-908

[13] Li Y, Chai F, Song Z, Li Z. Analysis of vibrations in interior permanent magnet synchro-
      nous motors considering air-gap deformation. Energies. 2017;**10**:1259

[14] Pfister P-D, Perriard Y. Very-high-speed slotless permanent-magnet motors: Analytical
      modeling, optimization, design, and torque measurement methods. IEEE Transactions
      on Industrial Electronics. 2010;**57**(1):296-303

[15] Luise F, Tessarolo A, Agnolet F, Pieri S, Scalabrin M, di Chiara M, de Martin M. Design
      optimization and testing of high-performance motors: Evaluating a compromise between
      quality design development and production costs of a Halbach-Array PM slotless motor.
      IEEE Industry Applications Magazine. 2016;**22**(6):19-32

[16] Krings A, Boglietti A, Cavagnino A, Sprague S. Soft magnetic material status and trends
      in electric machines. IEEE Transactions on Industrial Electronics. 2017;**64**(3):2405-2414

[17] Fan T, Li Q, Wen X. Development of a high power density motor made of amorphous alloy cores. IEEE Transactions on Industrial Electronics. 2014;**61**(9):4510-4518

[18] Okamoto S, Denis N, Kato Y, Ieki M, Fujisaki K. Core loss reduction of an interior permanent-magnet synchronous motor using amorphous stator core. IEEE Transactions on Industrial Applications. 2016;**52**(3):2261-2268

[19] Lei G, Zhu J, Guo Y, Liu C, Ma B. A review of design optimization methods for electrical machines. Energies. 2017;**10**(12):1962. DOI: 10.3390/en10121962

[20] Liu X, Slemon G. An improved method of optimization for electrical machines. IEEE Transactions on Energy Conversion. 1991;**6**(3):492-496

[21] Miller T. Optimal design of switched reluctance motors. IEEE Transactions on Industrial Electronics. 2002;**49**(1):15-27

[22] Andersen SB, Santos IF. Evolution strategies and multiobjective optimization of permanent magnet motor. Applied Soft Computing. 2012;**12**(2):778-792

[23] Zarko D, Ban D, Lipo T. Design optimization of interior permanent magnet (IPM) motors with maximized torque output in the entire speed range. In: European Conference on Power Electronics and Applications; 2005. 10pp

[24] Zhang P, Sizov G, Ionel D, Demerdash N. Establishing the relative merits of interior and spoke-type permanent magnet machines with ferrite or NdFeB through systematic design optimization. IEEE Transactions on Industry Applications. 2015;**99**:1

[25] Storn R, Price K. Differential Evolution—A simple and efficient adaptive scheme for global optimization over continuous spaces. Technical Report TR-95-012. ICSI; March 1995

[26] Lampinen J, Zelinka I. Mixed integer-discrete-continuous optimization by differential evolution—Part 1: The optimization method. In: 5th International Mendel Conference on Soft Computing; 1999. pp. 71-76

[27] Kukkonen S, Lampinen J. Performance assessment of generalized differential evolution 3 with a given set of constrained multi-objective test problems. In: IEEE Congress on Evolutionary Computation, CEC 09; 2009. pp. 1943-1950

[28] Lukaniszyn M, JagieLa M, Wrobel R. Optimization of permanent magnet shape for minimum cogging torque using a genetic algorithm. IEEE Transactions on Magnetics. 2004;**40**(2):1228-1231

[29] Bianchi N, Durello D, Fornasiero E. Multi-objective optimization of a PM assisted synchronous reluctance machine, including torque and sensorless detection capability. In: 6th IET International Conference on Power Electronics, Machines and Drives (PEMD 2012); March 2012. pp. 1-6

[30] Duan Y, Harley R, Habetler T. Multi-objective design optimization of surface mount permanent magnet machine with particle swarm intelligence. In: IEEE Swarm Intelligence Symposium; September 2008. pp. 1-5

[31] Ma C, Qu L. Multiobjective optimization of switched reluctance motors based on design of experiments and particle swarm optimization. IEEE Transactions on Energy Conversion. 2015;**99**:1-10

[32] Mutluer M, Bilgin O. Comparison of stochastic optimization methods for design optimization of permanent magnet synchronous motor. Neural Computing and Applications. 2012;**21**(8):2049-2056

[33] Deb A, Gupta B, Roy J. Performance comparison of differential evolution, genetic algorithm and particle swarm optimization in impedance matching of aperture coupled microstrip antennas. In: 11th Mediterranean Microwave Symposium (MMS); September 2011. pp. 17-20

[34] Duan Y, Harley R, Habetler T. Comparison of particle swarm optimization and genetic algorithm in the design of permanent magnet motors. In: IEEE 6th International Power Electronics and Motion Control Conference, IPEMC; 2009. pp. 822-825

[35] Mutluer M, Bilgin O. Design optimization of PMSM by particle swarm optimization and genetic algorithm. In: International Symposium on Innovations in Intelligent Systems and Applications (INISTA), 2012; July 2012. pp. 1-4

[36] Sarikhani A, Mohammed O. Hybrid GA-PSO multi-objective design optimization of coupled PM synchronous motor-drive using physics-based modeling approach. In: 14th Biennial IEEE Conference on Electromagnetic Field Computation (CEFC); 2010. pp. 1-1

[37] Stipetic S, Miebach W, Zarko D. Optimization in design of electric machines: Methodology and workflow. In: International Aegean Conference on Electrical Machines & Power Electronics (ACEMP); Side, Turkey. 2-4 September 2015. pp. 441-448

[38] Zhong C, Lin Y. Model reference adaptive control (MRAC)-based parameter identification applied to surface-mounted permanent magnet synchronous motor. International Journal of Electronics. 2017;**104**(11):1854-1873

[39] Esmaeeli MR, Kianinejad R, Razzaz M. Field oriented control of DFIG based on modified MRAS observer. In: Proceedings of 17th Conference on Electrical Power Distribution Networks (EPDC); Tehran, Iran. May 2012. pp. 2-3

[40] Zhang Y, Xia B, Yang H, Rodriguez J. Overview of model predictive control for induction motor drives. Chinese Journal of Electrical Engineering. 2016;**2**(1):62-76

[41] Hedjar R, Toumi R, Boucher P, Dumur D. Cascaded nonlinear predictive control of induction motor. European Journal of Control. 2004;**10**(1):65-80

[42] Moutchou M, Moahmoudi H, Abbou A. Backstepping control of the induction machine, based on flux sliding mode observer, with rotor and stator resistances adaptation. International Review of Automatic Control (IREACO). 2014;**7**(4):394-402

[43] Lin F-J, Shen P-H, Hsu S-P. Adaptive backstepping sliding mode control for linear induction motor drive. IEE Proceedings–Electric Power Applications. 2002;**149**(3):184-194

[44] Bolognani S, Zigliotto M. Fuzzy logic control of a switched reluctance motor drive. IEEE Transactions on Industry Applications. 1996;**32**(5):1063-1068

Multi-Objective Optimization Techniques to Solve the Economic Emission Load Dispatch Problem Using Various Heuristic and Metaheuristic Algorithms

Jorge de Almeida Brito Júnior,
Marcus Vinicius Alves Nunes,
Manoel Henrique Reis Nascimento,
Jandecy Cabral Leite,
Jorge Laureano Moya Rodriguez,
Carlos Alberto Oliveira de Freitas,
Milton Fonseca Júnior, Edson Farias de Oliveira,
David Barbosa de Alencar, Nadime Mustafa Moraes,
Tirso Lorenzo Reyes Carvajal and
Haroldo Melo de Oliveira

Additional information is available at the end of the chapter

http://dx.doi.org/10.5772/intechopen.76666

Abstract

The main objective of thermoelectric power plants is to meet the power demand with the lowest fuel cost and emission levels of pollutant and greenhouse gas emissions, considering the operational restrictions of the power plant. Optimization techniques have been widely used to solve engineering problems as in this case with the objective of minimizing the cost and the pollution damages. Heuristic and metaheuristic algorithms have been extensively studied and used to successfully solve this multi-objective problem. This chapter, several optimization techniques (simulated annealing, ant lion, dragonfly, NSGA II, and differential evolution) are analyzed and their application to economic-emission load dispatch (EELD) is also discussed. In addition, a comparison of all approaches and its results are offered through a case study.

Keywords: economic-emission load dispatch, optimization techniques, heuristic, metaheuristic algorithms, power plants

1. Introduction

The main problem of economic-emission load dispatch (EELD) is to reduce the emission level and total cost of generation at the same time accomplishing the demand for electricity from the power plant. Thermal power plants are among the maximum significant sources of contamination by sulfur dioxide (SO_2), carbon dioxide (CO_2), and nitrogen oxides (NOx), which create atmospheric pollution [1].

Some authors developed three approaches to solve the EELD problem [2–4]. The first one is using a single objective, considering emissions and pollution as restrictions with permissible limits [5]; the second one combines cost and emission functions into a single objective function with different weights, where cost and emission are minimized simultaneously [2]; and the third one uses the separated cost and emission functions in a multi-objective optimization [1, 6].

The solution of EELD problem is to minimize the total cost of fuel consumption and carbon emissions [7], considering power demand and operational restrictions [8]. Several techniques such as particle swarm optimization [9, 10], linear programming [11, 12], ant colony optimization [13], biogeography-based optimization [14], genetic algorithms (GA) [15, 16], Tabu search algorithm [17], simulated annealing (SA) [1], neural networks [18], differential evolution (DE) [19], harmony search algorithm [20], Lagrange functions [7], and others [19] have been used to fix the problem of EELD. In spite that all of them have been used, few are used with cost and emission functions in a multi-objective optimization.

This chapter, six multi-objective optimization approaches to solve the EELD problem are going to be presented, and a brief comparison among them is done.

In addition, the "shutdown" of the most inefficient generators is included in all multi-objective optimization approaches used [1, 8, 21].

2. Materials and methods

To solve a problem of EELD, two important objectives in an electrical thermal power system must be considered; they are environmental and economy impacts [22].

The parameters of objective functions are determined by curve fitting techniques based on tests of engine performance.

The multi-objective optimization problem is defined as:

$$\underset{P}{\text{minimize}} \quad [F_1(P), F_2(P)] \tag{1}$$

where F1(P) and F2(P) are the objective functions to be minimized over the set of permissible decision vector P, as follow in the next Subsections 2.1 and 2.2.

2.1. Cost function (F_1)

The fuel cost is considered as an essential criterion for economics analysis in thermal power plants. The cost function of each generator can be assumed to be approximated by a quadratic function of generator power output Pi [8, 22]:

$$F_1(P_i) = \sum_{i=1}^{n} \left(a_i + b_i P_i + c_i P_i^2\right) \quad \$/h$$

(2)

where a_i, b_i, and c_i are the fuel cost coefficients of the ith unit generating, n the number of generators, and P_i the active power of each generator (**Table 2**).

2.2. Economic emission function (F_2)

Emissions can be represented by a function that links emissions with power generated by each unit [23]. The emission function in ton/h, which normally represents the emission of SO_2 and NOx, is a function of the power output of the generator, and it can be expressed as follows [1]:

$$F_2(P_i) = \sum_{i=1}^{n} \left(d_i + e_i P_i + f_i P_i^2\right) \quad kg/h$$

(3)

where di, ei, and fi are emission coefficients.

2.3. Economical load dispatch constrains

The restrictions used in the problem were of three types as follows.

2.3.1. Equality power balance constrain

The real power of each generator is limited by the lower and upper limits. The following equation is the equality restriction of power balance [24, 25]:

$$\sum_{i=1}^{n} P_i - P^D - P^L = 0$$

(4)

where P_i is the output power of each i generator, P^D is the load demand, and P^L are transmission losses; in other words, the total power generation has to meet the total demand P^D and the actual power losses in transmission lines P^L, i.e.

$$\sum_{i=1}^{n} P_i = P^D + P^L$$

(5)

The calculation of power losses P^L involves the solution of the load flow problem, which has equality constraints in the active and reactive power on each bar as follows [26]:

$$P^L = \sum_{i=1}^{n} B_i P_i^2$$

(6)

A simplification is applied to model the transmission losses, setting them as a function of the generator output through Kron's loss coefficient derivatives of the Kron formula for losses [21]:

$$P_L = \sum_{i=1}^{n} \sum_{j=1}^{n} P_i B_{ij} P_j + \sum_{i=1}^{n} B_{0i} P_i + B_{00},$$

(7)

where B_{ij}, B_{0i} and B_{00} are the energy loss coefficients in the transmission network and n is the number of generators. A reasonable accuracy can be obtained when the actual operating conditions are close to the base case, where the B coefficients were obtained [26].

2.3.2. Production capacity constraint

The power capacity total generated from each generator is restricted by the lower limit and by the upper limit, so the constrain is [1]

$$P_{min.i} \leq P_i \leq P_{max.i}.$$

(8)

where P_i is the output power of the i generator, $P_{min.i}$ is the minimal power of the i generator, and $P_{max.i}$, the maximal power of the i generator.

2.3.3. Fuel delivery constraint

At each time interval, the amount of fuel supplied to all units must be less than or equal to the fuel supplied by the seller, i.e., the fuel delivered to each unit in each interval should be within its lower limit $F_{min,i}$ and its upper limit $F_{max,i}$ so that [21].

$$F_{min.i} \leq F_{im} \leq F_{max.i}, \; i \in N, m \in M,$$

(9)

where $F_{i,m}$ is the fuel supplied to the engine i at the interval m, $F_{i,min}$ is the minimum amount of fuel supplied to i generator, and $F_{max,i}$ is the maximum amount of fuel supplied to i generator.

3. Multi-objective optimization techniques to solve EELD

There are a lot of multi-objective optimization approaches that can be used to solve the EELD problem [19], but implementations need to be made to solve EELD problem. On this chapter six metaheuristic algorithm methods that already have the implementation are going to be presented.

3.1. Simulated annealing (SA)

SA is a powerful optimization technique which exploits the resemblance between a minimization process and the cooling of molten metal [1].

The physical annealing process is simulated in the SA technique for the determination of global or near-global optimum solutions for optimization problems. In this algorithm a parameter T, called temperature, is defined [27].

Algorithm 1 Simulated annealing

1: **procedure** SIMULATED-ANNEALING re-
    turns A STATE s_k
2:        **inputs**: T_0, initial temperature
3:                      J, cost function
4:                      s_0, initial state
5:                      *temp-schedule*, cooling schedule
6:                      *neighbor*, neighbor state function
7:      $T \leftarrow T_0$
8:      $s_k \leftarrow s_0$
9:      **for** $t = 1$ to $t_{\max}$ **do**
10:          $s_{k+1} \leftarrow neighbor(s_k)$
11:          $\Delta E \leftarrow J(s_{k+1}) - J(s_k)$
12:          **if** $\min(1, e^{-\Delta E/T}) \geqslant rand(0, 1)$ **then**
13:              $s_k \leftarrow s_{k+1}$
14:          **end if**
15:          $T \leftarrow$ temp-schedule(t)
16:      **end for**
17: **end procedure**

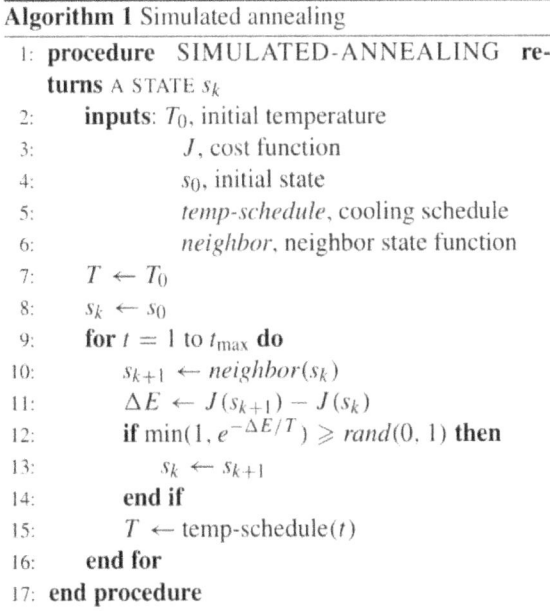

Figure 1. Simulated annealing algorithm. Source: [28].

On **Figure 1**, the simulated annealing algorithm is shown.

Starting from a high temperature, a molten metal is cooled slowly until it is solidified at a low temperature. The iteration number in the SA technique is analogous to the temperature level. In each iteration, a candidate solution is generated. If this solution is a better solution, it will be accepted and used to generate yet another candidate solution. If it is a deteriorated solution, the solution will be accepted when its probability of acceptance $Pr(\Delta)$ as given in Eq. (10) is greater than a randomly generated number between 0 and 1 [27]:

$$Pr(\Delta) = \exp(-\Delta/T) \tag{10}$$

where Δ is the amount of deterioration between the new and the current solutions and T is the temperature at which the new solution is generated. Accepting deteriorated solutions in the above manner enables the SA solution to "jump" out of the local optimum solution points and to seek the global optimum solution [1]. The last accepted candidate solution is then taken as the starting solution for the generation of candidate solutions in the next iteration. The reduction of the temperature in successive iterations is governed by the following geometric function:

$$T_v = r^{(v-1)}T_0 \tag{11}$$

where v is the iteration number and r is temperature reduction factor. T_0 is the initial temperature; its value can be set arbitrarily or estimated [29]. A multi-objective simulated annealing optimization to fix a generation dispatch problem is analyzed in [30], and used too in [1], but with the implementation of turning off the most inefficient generators, and this information will be used to compare with the new methods presented in this chapter.

3.2. NSGA II

The Non-dominated Sorting Genetic Algorithm II (NSGA II) has been one of the algorithms most used to solve the multi-objective optimization (MOO) problems in general and EELD problem in particular [31–33]. It was proposed by Deb et al. in 2002 [34]. NSGA II uses a faster selection, an elitist preservation approach, and a less segmented operating parameter [35]. The operation of NSGA II is shown in **Figure 2**.

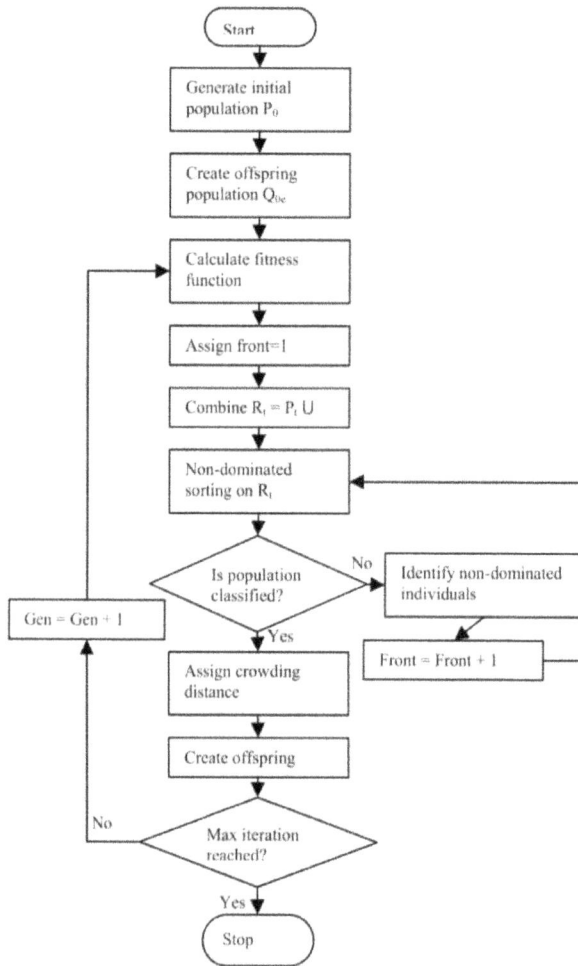

Figure 2. Operation of NSGA II. Source: [35].

3.2.1. Initial population

Initially, a parent random population is created. The population is ordered based on non-domination. For each solution, an aptitude equal to its level of non-domination is assigned (1 is the best level). Thus, a minimization of aptitude or fitness is assumed. Tournament selection, recombination, and mutation operators are used to create N size descendant populations [36].

The population is initialized based on the problem range and constraints, if any. This population is initialized and ordered based on non-domination.

3.2.2. Selection

In the original NSGA II, the selection by binary tournament (BTS) is used, where tournament is disputed between two solutions and the best one is selected and placed in the crossing tank. Two other solutions are taken again, and another gap in the cross junction is filled. It is performed in such a way that each solution can participate in two tournaments exactly [35].

3.2.3. Crossover

The NSGA II uses the simulated binary crossover (SBX), which works with two parent solutions and generates two descendants. The step-by-step procedure is described in [35].

3.2.4. Mutation

NSGA II uses polynomial mutation, which transforms each solution separately, that is, a stock solution gives offspring after being mutated. The mathematical formulation can be described as in [37].

3.2.5. Crowded tournament selection

An estimate obtained of the density of solutions close to a particular solution i in the population, the average between the two solutions on both sides of solution i together with each one of the targets is taken. This amount is the Crowding distance [35].

The NSGA II presents an improved technique for the maintenance of the diversity of solutions, proposing a method of crowding classification distance. However, in its case the classification technique remained unchanged from its previous version.

In NSGA II, all individuals are classified into combined populations (parents and descendants) based on the Pareto dominance ratio and then classified into several layers based on the front to which an individual is located. At each front, the individuals are arranged in descending order of magnitude of the crowding distance value. In the binary tournament selection process, the algorithm first selects an individual positioned on a better non-dominated front. In cases where the individuals with an identical front are compared, the tournament selection operator chooses the winner based on the crowding distance value [38].

The elitist strategy of NSGA II considers all combinations of non-dominated population solutions as the candidates for solutions to the next generation. If the number of non-dominated

solutions is less than the size of the population, they are all maintained as the next-genera-
tion solutions. Otherwise, candidates for the next generation are selected by the criterion of
crowding distance. This criterion has the advantage of maintaining the diversity of solutions
in the population in order to avoid premature convergence [38, 39].

3.3. Dragonfly

Recently new stochastic optimization technique was developed by Mirjaili in 2016 [40] named
dragonfly optimizer. Dragonflies (Odonata) are fancy insects. There are nearly 3000 different
species of this insect around the world. A dragonfly's lifecycle includes two main milestones:
nymph and adult. They spend the major portion of their lifespan in nymph, and they undergo
metamorphism to become adult [41].

On **Table 1** and **Figure 3**, the characteristics of dragonfly algorithm (DA) are shown.

Dragonflies are considered as small predators that hunt almost all other small insects in
nature. Nymph dragonflies also predate on other marine insects and even small fishes. The
interesting fact about dragonflies is their unique and rare swarming behavior. Dragonflies
swarm for only two purposes: hunting and migration. The former is called static (feeding)
swarm, and the latter is called dynamic (migratory) swarm [40].

The main inspiration of the DA algorithm originates from static and dynamic swarming
behaviors. These two swarming behaviors are very similar to the two main phases of optimization using meta-heuristics: exploration and exploitation. Dragonflies create sub-swarms
and fly over different areas in a static swarm, which is the main objective of the exploration
phase. In the static swarm, however, dragonflies fly in bigger swarms and along one direction,
which is favorable in the exploitation phase [40]. The dragonfly algorithm has been applied
successfully to the environmental economic dispatch.

General algorithm	Dragonfly algorithm
Decision variable	Dragonfly's position in each dimension
Solution	Dragonfly's position
Old solution	Old position of a dragonfly
New solution	New position of a dragonfly
Best solution	Any dragonfly with the best fitness
Fitness function	Distance from the food source, the predator, center of the swarm, velocity matching, and collision avoidance
Initial solution	Randomly selected position of a dragonfly
Selection	—
Process of generating new solution	Flying with a specific velocity and direction

Source: [42].

Table 1. Characteristics of the DA.

Algorithm 1 Pseudo-code of DA

1: initialize size of swarm N
2: initialize food source fitness F_{fit} and location F_{loc}
3: initialize enemy fitness E_{fit} and location E_{loc}
4: initialize number of neighbor J
5: initialize maximum iteration k_{max}
6: initialize objective function $func(X)$
7: **for** search agent i **to** N **do**
8:      initialize search agent location X_i
9:      initialize search agent step vector ΔX_i
10: **end for**
11: **for** iteration $k = 1$ **to** k_{max} **do**
12:     define w, s, a, c, f, e
13:     **for** each search agent i **do**
14:         calculate fitness value $func(X_i)$
15:         **if** $func(X_i)$ is better than F_{fit} **then**
16:             $F_{fit} \leftarrow func(X_i)$
17:             $F_{loc} \leftarrow X_i$
18:         **end if**
19:         **if** $func(X_i)$ is worse than E_{fit} **then**
20:             $E_{fit} \leftarrow func(X_i)$
21:             $E_{loc} \leftarrow X_i$
22:         **end if**
23:     **end for**
24:     **for** each search agent X_i **do**
25:         calculate J
26:         calculate S_i, A_i, C_i, F_i, E_i using Equation (1-5)
27:         **if** $J > 0$ **then**
28:             update ΔX_i using Equation (6)
29:             update X_i using Equation (7)
30:         **else**
31:             update X_i using Equation (8)
32:         **end if**
33:     **end for**
34: **end for**

Figure 3. Dragonfly pseudocode. Source: [43].

3.4. Particle swarm optimization (PSO)

Particle swarm optimization (PSO) is a population-based stochastic optimization technique, inspired by the social behavior of birds or shoals of fishes. Moore and Chapman extended this idea for multi-objective optimization in 1999 [44].

The simple PSO cannot be applied directly to multi-objective optimization since there are two issues to consider when extending PSO optimization to multi-objective problems [44].

The first one is how to select the best global and local particles (leaders) to guide the search, and the second one is how to keep good points found so far. For the latter, a secondary population is usually used to maintain non-dominated solutions.

PSO is one of the modern heuristic algorithms suitable for solving large-scale non-converting optimization problems. It is a population-based search algorithm and parallel search using a group of particles [45].

On **Figure 4**, the PSO flow chart is shown.

The main idea of the PSO is that "particles" (solutions) move through the search space with velocities that are adjusted dynamically according to their historical behavior. Therefore, the particles have a tendency to move to a better search area throughout the search process [47].

3.5. Differential evolution (DE)

The DE algorithm was developed to be a promising heuristic optimization algorithm for numerical optimization problems. The DE was designed to meet the requirement of practical minimization techniques with consistent convergence to the global minimum in consecutive independent trials. It can solve non-differentiable, nonlinear functions and multimodal cost functions [48].

The DE algorithm is a simple stochastic optimization strategy. It uses a voracious and less stochastic approach with floating-point coding in problem solving, unlike other evolutionary algorithms [49].

DE uses the arithmetic operators to evolve from a randomly generated initial population to a final solution. Basically, the weighted difference between two individuals is added to a third individual in the population. Thus, no separate probability distribution has to be used, which makes the scheme completely self-organized [49]. There are several variant strategies of DE. In general expressions are divided into two parts: the first part represents a vector to be disturbed. The first part is "rand" (vector chosen at random) or "best" (best vector of the current population). The second part is the number of difference vectors (one or two) chosen for perturbation vectors of the first part, and the last part indicates the type of crossover to be used. The type of crossing can be "bin" (binomial) or "exp" (exponential) [49].

3.5.1. Initialization

The initialization of DE can be done by randomly generating candidate solutions with NP D-dimensional vectors of real parameters evaluated [50].

There are other approaches to determining the initial population, although random equality is the most common. In current optimization problems, a possible solution parameter is added to the initial population in order to improve convergence.

3.5.2. Mutation

Differential mutation adds a scale, random sampling, of a difference vector to a third vector. Mutant vectors, also called donors, are obtained through the differential mutation operation [50].

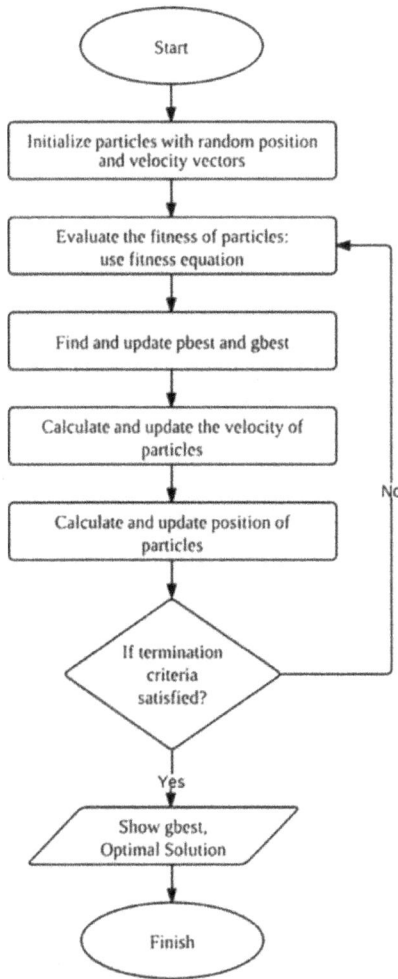

Figure 4. PSO flow chart. Source: [46].

3.5.3. Crossover

The crossover increases the potential diversity of a population. In the case of binomial crossover, test vectors are produced. Crossover can be understood as a mutation rate or a probability of inheritance between successive generations. There are alternatives to binomial crossover. The most common is the exponential crossover. Both approaches are valid for all problems, although the success/improvement of one over the other varies according to the problem considered [50].

3.5.4. Selection

Selection can be understood as a form of competition, in line with many examples directly observable in nature. Many evolutionary optimization schemes, such as DE or genetic algorithms (GAs), use some form of selection.

For a selection operation, the selection of pairs, also called avid selection or elite selection, is constantly used in the algorithm. As a stopping criterion, a maximum number of generations is defined [50].

3.6. Ant lion

Ant lion optimizer (ALO) [51] is a new algorithm inspired by nature proposed by Seyedali Mirjalili in 2015. The ALO algorithm mimics the mechanism of ant hunting in nature. It uses five main stages of prey hunting, such as random walking of ants, building traps, trapping ants in traps, catching prey, and rebuilding traps. All these steps are implemented.

The ant lions belong to the class of insects with wings and nerves (Neuroptera). The life cycle of the ants includes two main phases: larvae and adults. A natural shelf life can take up to 3 years, which occurs mainly in larvae (3–5 weeks into adulthood). The ant lion undergoes a metamorphosis into a cocoon to become an adult.

They hunt primarily on larvae, and the adult period is for breeding. A larva of ant lion digs a well of cones into sand, moving along a circular path and throwing the sands with the massive jaw. After digging the trap, the larvae hide under the bottom of the cone and wait for the insects (preferably ants) to be trapped in the well. The edge of the cone is sharp enough so that the insects fall easily into the bottom of the trap. Once that ALO realizes that a prey is in the trap, it tries to catch it. This is one of the algorithms that are also used for EELD, and it is one of the most recent discoveries [52–54].

The ALO is governed by the following rules [51]:

1. Ants move around the search space using different random walks.

2. Random walks are affected by the traps of ant lions.

3. Ant lions can build pits proportional to their fitness (the higher the fitness, the larger the pit).

4. Size of the pits is proportional to the probability of catching prey. Hence, ant lions with larger pits have higher probability to catch ants.

5. Each ant can be caught by an ant lion as well as the elite (fittest ant lion) in each iteration.

6. When ants try to escape from the pit, the ant lions throw sand toward the top of the trap to slide the ants inside the bottom of the trap. In order to simulate this behavior of sliding ants toward ant lions, the range of random walk is decreased adaptively.

7. If an ant becomes fitter than an ant lion, this means that it is caught and pulled under the sand by the ant lion.

8. After each hunt, an ant lion repositions itself to the latest caught prey and builds a pit to improve its chance of catching another prey.

4. Results of the comparison among all approaches applied to a case study

The power plant selected for the case study consists of 10 gas engines Jenbacher J620. In order to take the data of the power plant, its 10 motors were placed to work at different powers

(20, 30, 40, 50, 60, 70, 80, and 100% of the total power). For each one of these regimes, the data of fuel consumption of each engine were taken, and then the curve of power vs. cost of fuel for each motor was plotted. For clarity is convenient to say that each consume data was measured 10 times and was chosen the main. In a similar way, it was done for the emission case. The engines were set to work with different power values, the data of the emissions of NO2, CO, CO2, SO2, etc., were collected, besides the total volume of the emissions of each engine, and the curves of power vs. emissions were also plotted. In this section, comparisons of results obtained with different approaches for the same case study are presented. There are many algorithms used for the economic environmental load dispatch [55]. Results are compared using the following algorithms:

- NSGA II

- Dragonfly optimizer

- Particle swarm optimization

- Simulated annealing

- Ant lion optimizer

The comparison was made with the following data:

- Power demand, 20 MW

- Number of power plant generators, 10

- Minimum power of generators, 0.56 MW

- Maximum power of generators, 3.9 MW

All algorithms were programmed in MATLAB to simulate and compare them with the pre-defined case studies. Besides, the shutdown of less efficient motors in all these metaheuristic algorithms was implemented.

4.1. Motor data

Table 3 shows the emission coefficients of 10 gas engines of the power plant used as a case study.

The coefficients "a," "b," "c," "d," "e," and "f" were determined by operating the engines of the power plant at different powers, measuring the consumption and emissions to obtain the power versus cost and power vs. emission curves of each engine. The quadratic equation of each curve of each motor was obtained by the regression methods using MATLAB's tool box curve fitting. In the same way, the coefficients "d," "e," and "f" were obtained but in this case by measuring the CO_2, NO_x, and SO_2 emissions of each engine at different powers.

On **Table 4**, the coefficients of losses of the ten motors that make up the case study are presented. In this chapter a reduction to the transmission loss model is applied as a function of the output of the generators through the Kron loss coefficients [26].

On **Table 5** cost and emission results with all metaheuristic algorithms used for comparison in this chapter are shown. Among these algorithms, applied in a plant with ten generators,

Motor	c_i ($/Mw^2)	b_i ($/Mw)	a_i($)	P_{min}(Mw)	P_{max} (Mw)
1	0.007	7	240	0.66	3.35
2	0.0095	10	200	0.9	3.7
3	0.009	8.5	220	0.8	3.6
4	0.009	11	200	0.66	3.35
5	0.008	10.5	220	0.72	3.45
6	0.0075	12	120	0.66	2.97
7	0.0075	14	130	0.88	3.5
8	0.0075	14	130	0.754	3.33
9	0.0075	14	130	0.9	3.9
10	0.0075	14	130	0.56	2.35

Source: [1, 21].

Table 2. Cost coefficients from the thermal plant case study.

Motor	f	e	d
1	0.00419	1.32767	73.85932
2	0.00419	0.32767	13.85932
3	0.00683	−0.54551	40.2669
4	0.00683	−0.54551	40.2669
5	0.00461	−0.51116	42.89553
6	0.00461	−0.51116	42.8955
7	0.00461	−0.51116	42.8955
8	0.00461	−0.51116	42.8955
9	0.00061	−0.51116	10.8955
10	0.00461	−0.51116	42.8955

Source: [1].

Table 3. Emission coefficients from the thermal power plant used as case study.

it is possible to notice that the SA has the lowest emission of pollutants with 1757.39 (m³/h), however with a cost of 1556.61 ($/h), while the DE algorithm has the lowest cost with 1545.59 ($/h), however with emission of pollutants with 1769.48 (m³/h).

The comparison of all the results of the different metaheuristic algorithms is provided in **Table 6**.

M	1	2	3	4	5	6	7	8	9	10
1	0.14	0.17	0.15	0.19	0.26	0.22	0.34	0.38	0.43	0.45
2	0.17	0.6	0.13	0.16	0.15	0.2	0.23	0.56	0.23	0.51
3	0.15	0.13	0.65	0.17	0.24	0.19	0.25	0.38	0.43	0.45
4	0.19	0.16	0.17	0.71	0.3	0.25	0.43	0.56	0.23	0.51
5	0.26	0.15	0.24	0.3	0.69	0.32	0.18	0.37	0.42	0.48
6	0.22	0.2	0.19	0.25	0.32	0.85	0.97	0.55	0.27	0.58
7	0.22	0.2	0.19	0.25	0.32	0.85	0.67	0.38	0.43	0.45
8	0.19	0.7	0.13	0.18	0.16	0.21	0.28	0.56	0.23	0.51
9	0.26	0.15	0.24	0.3	0.69	0.32	0.18	0.37	0.42	0.48
10	0.15	0.13	0.65	0.17	0.24	0.19	0.25	0.38	0.43	0.45

Fonte: [1, 21].

Table 4. Loss coefficients (all values have to be multiplied by 1e-4).

Results	SA	DE	DA	NSGA II	PSO	ALO
Emission (m3/h)	1757,39	1769,48	1765,34	2098,12	2113,33	1763,55
Cost ($/h)	1556,61	1545,59	1548,28	1709,47	1685,18	1549,93

Table 5. Comparative table with all costs and emissions.

Method		Motor 1	Motor 2	Motor 3	Motor 4	Motor 5	Motor 6	Motor 7	Motor 8	Motor 9	Motor 10	Total
Simulates Anneling	Power (MW)	1,98	2,06	3,6	3,35	3,44	0	0	3,27	0	2,31	20,01
	Emission (m³/h)	100,98	101,08	288,12	288,67	308,27	0	0	318,88	0	351,39	1757,39
	Cost ($)	253,91	220,65	250,72	236,95	256,27	0	0	175,79	0	162,32	1556,61
Differential Evolution	Power (MW)	3,32	2,49	3,59	3,35	3,44	0	0	1,71	0	2,1	20,01
	Emission (m³/h)	106,66	102,13	288,13	288,67	308,27	0	0	323,55	0	352,06	1769,48
	Cost ($)	263,29	224,98	250,65	236,94	256,26	0	0	154,02	0	159,44	1545,59
NSGA II	Power (MW)	0,66	1,11	3,53	3,24	3,35	0	0	2,77	3,71	2,26	20,63
	Emission (m³/h)	101,87	101,1	288,27	288,95	308,52	0	0	320,18	337,71	351,52	2098,12
	Cost ($)	244,62	211,13	250,09	235,69	255,25	0	0	168,86	182,09	161,73	1709,47
Particle Swarm	Power (MW)	2,64	2,56	3,33	2,89	2,78	0	0	1,45	3,68	0,69	20,01
	Emission (m³/h)	102,67	102,38	288,72	289,87	310,15	0	0	324,5	337,78	357,28	2113,33
	Cost ($)	258,51	225,68	248,4	231,83	249,2	0	0	150,28	181,64	139,63	1685,18
Dragonfly	Power (MW)	3,04	2,28	3,6	3,35	3,45	0	0	2,21	0	2,08	20,01
	Emission (m³/h)	104,73	101,52	288,12	288,67	308,26	0	0	321,89	0	352,15	1765,34
	Cost ($)	261,35	222,89	250,72	236,95	256,32	0	0	160,97	0	159,09	1548,28
Ant Lion	Power (MW)	2,81	2,33	3,59	3,33	3,43	0	0	2,59	0	1,94	20,01
	Emission (m³/h)	103,42	101,64	288,13	288,72	308,31	0	0	320,71	0	352,62	1763,55
	Cost ($)	259,7	223,33	250,64	236,72	256,11	0	0	166,29	0	157,15	1549,93

Table 6. Comparison between all results of the different programmed algorithms.

5. Conclusion

The model was implemented in MATLAB computing. Usually all algorithms have presented good results, because in all cases it is possible to switch off at least two motors, but some differences among results of the different algorithms were evidenced. It was possible to notice vantages in relation to cost and emission of two methods among all of them. The SA achieves the lowest emission of pollutants, while DE obtained the lowest cost for the power plant of ten generating units.

Acknowledgements

The Institute of Technology and Education "Galileo" from Amazonia (ITEGAM), the Federal University of Para (UFPA), the Research Support Foundation State of Amazonas (FAPEAM), and the National Council of Research (CNPq) Productivity of Research Funds Process 301105/2016-2 for the financial support to this research.

Author details

Jorge de Almeida Brito Júnior[1]*, Marcus Vinicius Alves Nunes[2],
Manoel Henrique Reis Nascimento[1], Jandecy Cabral Leite[1],
Jorge Laureano Moya Rodriguez[3], Carlos Alberto Oliveira de Freitas[1],
Milton Fonseca Júnior[4], Edson Farias de Oliveira[1], David Barbosa de Alencar[1],
Nadime Mustafa Moraes[5], Tirso Lorenzo Reyes Carvajal[1] and Haroldo Melo de Oliveira[1]

*Address all correspondence to: jorge.brito@itegam.org.br

1 Research Department Institute of Technology and Education Galileo da Amazônia (ITEGAM), Manaus, Brazil

2 Faculty of Electrical Engineering Institute of Technology, Federal University of Para (UFPA), Belém, Pará, Brazil

3 Federal University of Bahia (UFBA), Bahia, Brazil

4 Department of Generation of Mauá, Eletrobrás Amazonas GT, Manaus, Amazonas, Brazil

5 University of the State of Amazonas (UEA), Manaus, Amazonas, Brazil

References

[1] Júnior JAB, Nunes MVA, Nascimento MHR, et al. Solution to economic emission load dispatch by simulated annealing: Case study. Electrical Engineering. 2017. https://doi.org/10.1007/s00202-017-0544-0

[2] Zhou J, Wang C, Li Y, Wang P, Li C, Lu P, et al. A multi-objective multi-population ant colony optimization for economic emission dispatch considering power system security. Applied Mathematical Modelling. 2017;**45**:684-704

[3] Roy PK, Bhui S. A multi-objective hybrid evolutionary algorithm for dynamic economic emission load dispatch. International Transactions on Electrical Energy Systems. 2016;**26**(1):49-78

[4] Modiri-Delshad M, Kaboli SHA, Taslimi-Renani E, Rahim NA. Backtracking search algorithm for solving economic dispatch problems with valve-point effects and multiple fuel options. Energy. 2016;**116**:637-649

[5] Granelli G, Montagna M, Pasini G, Marannino P. Emission constrained dynamic dispatch. Electric Power Systems Research. 1992;**24**(1):55-64

[6] Gonçalves E. Métodos híbridos de pontos interiores/exteriores e de aproximantes de funções em problemas multiobjetivo de despacho econômico e ambiental; 2015

[7] Krishnamurthy S, Tzoneva R, editors. Comparative analyses of Min-Max and Max-Max price penalty factor approaches for multi criteria power system dispatch problem with valve point effect loading using Lagrange's method. 2011 International Conference on Power and Energy Systems; 2011:22-24 Dec. 2011

[8] Nascimento MHR, Nunes MVA, Rodríguez JLM, Leite JC, Junior JAB. New solution for resolution of the economic load dispatch by different mathematical optimization methods, turning off the less efficient generators. Journal of Engineering and Technology for Industrial Applications. 2017;**03**:37-46

[9] De M, Das G, Mandal S, Mandal K. Investigating economic emission dispatch problem using improved particle swarm optimization technique. Industry Interactive Innovations in Science, Engineering and Technology: Springer. 2018:37-45

[10] Mason K, Duggan J, Howley E. Multi-objective dynamic economic emission dispatch using particle swarm optimisation variants. Neurocomputing. 2017;**270**(Suppl C):188-197

[11] Mohan M, Kuppusamy K, Khan MA. Optimal short-term hydrothermal scheduling using decomposition approach and linear programming method. International Journal of Electrical Power & Energy Systems. 1992;**14**(1):39-44

[12] Farag A, Al-Baiyat S, Cheng T. Economic load dispatch multiobjective optimization procedures using linear programming techniques. IEEE Transactions on Power Systems. 1995;**10**(2):731-738

[13] Huang S-J. Enhancement of hydroelectric generation scheduling using ant colony system based optimization approaches. IEEE Transactions on Energy Conversion. 2001;**16**(3):296-301

[14] Ma H, Yang Z, You P, Fei M. Multi-objective biogeography-based optimization for dynamic economic emission load dispatch considering plug-in electric vehicles charging. Energy. 2017;**135**(Suppl C):101-111

[15] Liu Hd, Ma Zl, Liu S, Lan H, editors. A new solution to economic emission load dispatch using immune genetic algorithm. 2006 IEEE Conference on Cybernetics and Intelligent Systems; 2006 7-9 June 2006

[16] Damousis IG, Bakirtzis AG, Dokopoulos PS. Network-constrained economic dispatch using real-coded genetic algorithm. IEEE Transactions on Power Systems. 2003;**18**(1):198-205

[17] Kumarappan N, Mohan MR, editors. Hybrid genetic algorithm based combined economic and emission dispatch for utility system. International Conference on Intelligent Sensing and Information Processing, 2004 Proceedings of; 2004

[18] Momoh JA, Reddy SS, editors. Combined Economic and Emission Dispatch using Radial Basis Function. 2014 IEEE PES General Meeting | Conference & Exposition; 2014 27-31 July 2014

[19] Jebaraj L, Venkatesan C, Soubache I. Rajan CCA. Application of differential evolution algorithm in static and dynamic economic or emission dispatch problem: A review. Renewable and Sustainable Energy Reviews. 2017;**77**(Suppl C):1206-1220

[20] Chatterjee A, Ghoshal S, Mukherjee V. Solution of combined economic and emission dispatch problems of power systems by an opposition-based harmony search algorithm. International Journal of Electrical Power & Energy Systems. 2012;**39**(1):9-20

[21] Nascimento MHR, Nunes MVA, Rodríguez JLM, et al. A new solution to the economical load dispatch of power plants and optimization using differential evolution. Electrical Engineering. 2017;**99**:561. https://doi.org/10.1007/s00202-016-0385-2

[22] Basu M. Fuel constrained economic emission dispatch using nondominated sorting genetic algorithm-II. Energy. 2014;**78**:649-664

[23] Miranda V, Hang PS. Economic dispatch model with fuzzy wind constraints and attitudes of dispatchers. IEEE Transactions on Power Systems. 2005;**20**(4):2143-2145

[24] Nwulu NI, Xia X. Multi-objective dynamic economic emission dispatch of electric power generation integrated with game theory based demand response programs. Energy Conversion and Management 2015;**89**(0):963-974

[25] Ashish D, Arunesh D, Surya P, Bhardwaj AK. A traditional approach to solve economic load dispatch problem of thermal generating unit using MATLAB programming. International Journal of Engineering Research & Technology (IJERT). 2013;**2**(9):2013

[26] Wang L, Singh C. Environmental/economic power dispatch using a fuzzified multi-objective particle swarm optimization algorithm. Electric Power Systems Research. 2007;**77**(12):1654-1664

[27] Kamboj VK, Bath S, Dhillon J. Solution of non-convex economic load dispatch problem using Grey wolf optimizer. Neural Computing and Applications. 2015:1-16

[28] Fraga-Gonzalez LF, Fuentes-Aguilar RQ, Garcia-Gonzalez A, Sanchez-Ante G. Adaptive simulated annealing for tuning PID controllers. AI Communications. 2017;**30**(5):347-362

[29] Wong K, Fung C, editors. Simulated annealing based economic dispatch algorithm. IEE proceedings C (generation, transmission and distribution). IET; 1993

[30] Ziane I, Benhamida F, Amel G. Simulated annealing optimization for multi-objective economic dispatch solution. Leonardo Journal of Sciences. 2014;13(25):43-56

[31] Abul'Wafa AR. Optimization of economic/emission load dispatch for hybrid generating systems using controlled elitist NSGA-II. Electric Power Systems Research. 2013;105:142-151

[32] Basu M. Combined heat and power economic emission dispatch using nondominated sorting genetic algorithm-II. International Journal of Electrical Power & Energy Systems. 2013;53:135-141

[33] Cococcioni M, Lazzerini B, Marcelloni F, Pistolesi F, editors. Solving the environmental economic dispatch problem with prohibited operating zones in microgrids using NSGA-II and TOPSIS. Proceedings of the 31st Annual ACM Symposium on Applied Computing. ACM; 2016

[34] Deb K, Pratap A, Agarwal S, Meyarivan T. A fast and elitist multiobjective genetic algorithm: NSGA-II. IEEE Transactions on Evolutionary Computation. 2002;6(2):182-197

[35] Golchha A, Qureshi SG. Non-dominated sorting genetic algorithm-II – A succinct survey. International Journal of Computer Science and Information Technologies - (IJCSIT). 2015;6(1):252-255

[36] Kalaivani L, Subburaj P, Willjuice Iruthayarajan M. Speed control of switched reluctance motor with torque ripple reduction using non-dominated sorting genetic algorithm (NSGA-II). International Journal of Electrical Power & Energy Systems. 2013;53:69-77

[37] Golchha A, Qureshi SG. Non-dominated sorting genetic algorithm-II–A succinct survey. International Journal of Computer Science and Information Technologies. 2015;6(1):252-255

[38] Suksonghong K, Boonlong K, Goh K-L. Multi-objective genetic algorithms for solving portfolio optimization problems in the electricity Market. Electrical Power and Energy Systems, Elsevier Ltd All rights reserved; 2014

[39] Liu T, Gao X, Wang L. Multi-objective optimization method using an improved NSGA-II algorithm for oil–gas production process. Journal of the Taiwan Institute of Chemical Engineers 2015(0)

[40] Mirjalili S. Dragonfly algorithm: A new meta-heuristic optimization technique for solving single-objective, discrete, and multi-objective problems. Neural Computing and Applications. 2016;27(4):1053-1073

[41] Zolghadr-Asli B, Bozorg-Haddad O, Chu X. Dragonfly Algorithm (DA). Advanced Optimization by Nature-Inspired Algorithms. Springer; 2018. pp. 151-159

[42] Bozorg-Haddad O. Advanced Optimization by Nature-Inspired Algorithms. Springer; 2018

[43] Daely PT, Shin SY, editors. Range based wireless node localization using dragonfly algorithm. Ubiquitous and Future Networks (ICUFN), 2016 Eighth International Conference on. IEEE; 2016

[44] Zhoua A, Bo-Yang Q, Shi-Zheng HLZ, Nagaratnam SP, Qingfu Z. Multiobjective evolutionary algorithms: A survey of the state of the art. Elsevier BV All rights reserved, Swarm and Evolutionary Computation; 2011

[45] Park JB, Jeong YW, Shin JR, Lee KY. An improved particle swarm optimization for nonconvex economic dispatch problems. IEEE Transactions on Power Systems. 2010;**25**(1):156-166

[46] Armaghani DJ, Hajihassani M, Mohamad ET, Marto A, Noorani S. Blasting-induced flyrock and ground vibration prediction through an expert artificial neural network based on particle swarm optimization. Arabian Journal of Geosciences. 2014;**7**(12):5383-5396

[47] Idoumghar L, Chérin N, Siarry P, Roche R, Miraoui A. Hybrid ICA–PSO algorithm for continuous optimization. Applied Mathematics and Computation. 2013;**219**(24):11149-11170

[48] Behera S, Sahoo S, Pati BB. A review on optimization algorithms and application to wind energy integration to grid. Renewable and Sustainable Energy Reviews. 2015;**48**:214-227

[49] Coelho LS, Bora TC, Mariani VC. Differential evolution based on truncated Lévy-type flights and population diversity measure to solve economic load dispatch problems. International Journal of Electrical Power & Energy Systems. 2014;**57**:178-188

[50] Roque CMC, Martins PALS. Differential evolution for optimization of functionally graded beams. Composite Structures. 2015;**133**:1191-1197

[51] Mirjalili S. The ant lion optimizer. Advances in Engineering Software. 2015;**83**:80-98

[52] Raju M, Saikia LC, Sinha N. Automatic generation control of a multi-area system using ant lion optimizer algorithm based PID plus second order derivative controller. International Journal of Electrical Power & Energy Systems. 2016;**80**:52-63

[53] Kamboj VK, Bhadoria A, Bath S. Solution of non-convex economic load dispatch problem for small-scale power systems using ant lion optimizer. Neural Computing and Applications. 2017;**28**(8):2181-2192

[54] Mirjalili S, Jangir P, Saremi S. Multi-objective ant lion optimizer: A multi-objective optimization algorithm for solving engineering problems. Applied Intelligence. 2017;**46**(1): 79-95

[55] Qu B, Zhu Y, Jiao Y, Wu M, Suganthan P, Liang J. A survey on multi-objective evolutionary algorithms for the solution of the environmental/economic dispatch problems. Swarm and Evolutionary Computation;**2017**

Chapter 3

Optimal Design of Brushless Doubly Fed Reluctance Machine

Mandar Bhawalkar, Gopalakrishnan Narayan and
Yogesh Nerkar

Additional information is available at the end of the chapter

http://dx.doi.org/10.5772/intechopen.74805

Abstract

Optimization techniques are widely used in the design of electrical machines to obtain maximum performance at minimal capital cost. After a brief overview of some of the optimization techniques employed in electrical machine design, this chapter highlights the features of brushless doubly fed reluctance machine (BDFRM) and its optimal design. The simple and robust construction, variable speed operation, better performance compared to traditional counterpart, and requirement of partially rated converter for speed control have made BDFRM an attractive alternative for variable speed applications such as pumps, blower, and wind generators. Due to unusual construction of BDFRM, conventional design procedures cannot be applied. A few critical issues in the design of BDFRM that greatly affect its performance are discussed. Design optimization is performed using nonlinear programming technique for 6-4-2 pole reluctance rotor and 8-6-4 pole ducted rotor configurations of BDFRM. 2 kW prototypes are then constructed for laboratory use. The performance of the prototypes is examined through finite element analysis (FEA) employing Maxwell 16 software. The test results are also presented.

Keywords: optimization, nonlinear programming, objective function, BDFRM, optimal design of BDFRM, finite element analysis

1. Introduction

Many industrial applications demand efficient, robust, and cost-effective variable speed drives. Brushless doubly fed reluctance machine (BDFRM) is one of the better alternatives for induction motors due to simple and rugged construction, higher efficiency, absence of rotor winding, slip rings and brushes, and reduced maintenance cost. Last but not least, a partially rated converter is sufficient for control of speed in either direction of rated speed [1, 2]. These benefits are useful to BDFRM in generator mode in wind power applications.

The fierce competition in world market, increasing cost of energy supplies, and regulations for conservation of energy and resources are the key forces for design optimization of electrical machines. Electrical machine design is a complex process which can be realized as articulation or formulation subjected to restrictions imposed by various factors such as nonlinear characteristic of different material compositions, performance parameters, etc. and ability to achieve acceptable performance. The expected outcome of optimization is to minimize the capital and running costs and increased service life [3, 4]. The optimization is aimed at getting desirable solution of objective function(s) to maximize performance subjected to constraints. The design by analysis is based on estimation of performance of a machine from known variables and their inter-dependability. The known variables are usually design specifications or system parameters. In the design, the desired performance parameters are substituted in realistic mathematical model of machine which is then solved iteratively. However, the range of acceptable solutions may not necessarily yield optimum results with respect to cost, material requirements, and other factors, and hence machine design problem is linked to optimization and needs to be solved iteratively.

The organization of this chapter is as follows: Section 2 presents an overview of optimization techniques, and Section 3 gives a brief introduction to BDFRM to develop better understanding of machine. Section 4 discusses design considerations of BDFRM. An objective function is developed to minimize the active material requirements. Nonlinear programming is carried out for the optimal design of BDFRM. Section 5 presents the development of 2 kW prototypes. Section 6 discusses simulation results using Maxwell 16 (finite element software). This followed by test results of prototypes in Section 7. Section 8 presents conclusions.

2. Overview of optimization techniques

The optimization problems can be classified [5] as shown in **Table 1**. The classification of optimization problem depends on the nature of problem, operational constraints, and design limitations. The selection of algorithm for solving optimal problem depends not only on the types of objective function used but also on how its first and second derivatives are computed.

The optimization methods can be further classified broadly as classical methods and recent or advanced methods. The classical methods use continuous and differentiable functions [5], and these methods include single variable and multivariable optimization with and without constraints. Depending on the nature of objective function and design variables, optimization problem can be linear or nonlinear optimization with or without constraints. There is an inexhaustive list of different optimization techniques used for different applications looking into the aspects of various design/operational/environmental constraints. These methods are

Based on	Classification
Existence of constraints	Unconstrained/constrained optimization
Nature of design variables	Static/dynamic/trajectory optimization
Physical structure of problem	Optimal/nonoptimal control
Equations involved and objective function	Linear/nonlinear/geometric/quadratic programming
Permissible values of the design variables	Integer programming/real-valued programming
Nature of variables	Deterministic/stochastic programming
Separable functions	Separable/non-separable programming
Number of objective functions	Single objective/multiobjective programming

Table 1. Classification of optimization problems.

Optimization	Optimization methods
Classical	Single variable, multivariable with or without constraints
LP	Simplex, revised simplex, dual simplex method
One-dimensional nonlinear programing (NLP)	Elimination methods—unrestricted search, exhaustive search, dichotomous search, Fibonacci method, golden section search Interpolation methods—quadratic interpolation, cubic interpolation Direct root methods—Newton, quasi-Newton, and Secant
NLP with no constraint	Direct search—random search, grid search, univariate, pattern search, Powell, Hooke-Jeeves, Rosenbrock, and simplex Descent methods—steepest descent, Fletcher-Reeves, Newton, Marquardt, quasi-Newton, Davidon-Fletcher-Powell, and Broyden-Fletcher-Goldfarb-Shanno
Constrained NLP	Direct methods—random search, heuristic search, complex, objective and constraint approximation, sequential linear programming, sequential quadratic programming, method of feasible directions, Zoutendijk, Rosen's gradient projection, generalized reduced gradient Indirect methods—transformation of variables, sequential unconstrained minimization, interior penalty, exterior penalty, and augmented Lagrange multiplier
Advanced optimization methods	Geometric programming, dynamic programming, integer programming, stochastic programming, simulated annealing method, genetic algorithm, artificial neural network, fuzzy system, particle swarm optimization, differential algorithm, filled function method, evolutionary algorithms, design of experiments—central composite design, Latin hypercube design

Table 2. Different optimization techniques.

given in **Table 2** [3–9]. Classical optimization methods are found wanting due to increasing complexities, inter-dependability of design variables, and constraint functions. Therefore, nonlinear optimization methods with constraints are widely used along with newer methods such as genetic algorithms, artificial neural networks, and fuzzy logic techniques.

In the optimal design of electrical machines, the model of problem to be solved is to be selected along with choice of algorithms, variables, constraint functions, and objective functions. The classical methods such as random search method, simplex method, Hooke and Jeeves method, Powell and David-Fletcher-Powell method, and penalty function methods have been used in earlier decades. Most of them are nonlinear programing types as nonlinearities are imposed due to materials, operational issues, etc. Mathematically, these problems can be stated as.

$$minimize\, f(X) \tag{1}$$

$$subjected\ to\ g(X) \geq 0 \tag{2}$$

where X is an n-dimensional vector, f(X) is an objective function, and g(X) represents design constraints. The bounded region enclosed by g(X) = 0 where X is feasible.

The multiobjective function optimization problems are more complex as the feasible solutions are conflicting and practically impossible to obtain optimal solution to all objective functions simultaneously. Hence, a set solutions providing tradeoff between objective functions are acceptable.

The recent methods such as differential evolutions (DE), genetic algorithms (GA), evolutionary algorithms (EA), and particle swarm optimization (PSO) are used in machine design optimization [7–9]. These methods use a set of populations as initial approximation unlike classical method and are stochastic in nature. As these methods can optimize nonlinear functions containing continuous and discrete variables, they are the most suited ones for machine design problems. A recent study [7] has shown that DE may not be always the fastest, but produces the best results. GA is based on principles of natural genetics and natural selection. It encodes parameters as bit string, and then manipulation is performed using logical operators [10]. Another technique design of experiments (DoE) is a statistical method that effectively quantifies the effect of changes in design variables or machine response. This method includes one factor at a time and is suitable for a limited number of variables. Hence, advanced DoE methods such as central composite design (CCD) and Latin hypercube design (LHD) are used [7]. EA encodes parameters as floating point arithmetic operators. These methods are usually linked to finite element analysis solvers to optimize the machine design. PSO is also gaining importance in electrical machine design as it is used to get appropriate fitness evaluation function, which represents the relationship between design variables and machine response. In PSO the potential solution (particles) flies through the problem space by following the current optimum particles after initialization with random particles. In addition ANN and fuzzy logic are also gaining wide-scale acceptance. ANN is used due its immense computational powers. Appropriate selection of network topology and layers is very important. This method also needs voluminous data for training neural networks. ANN can be trained to learn the relationship between input and output parameters of electrical machines [11]. Fuzzy logic is also being tried out for electrical machine design [12]. The fuzzy sets are formed from decision and controlled variables. A qualitative approach is used for deriving membership and objective functions. A combination of stochastic method such as such as neuro-fuzzy, fuzzy, and genetic algorithm is also used in optimization [13].

3. Brushless doubly fed reluctance machine (BDFRM)

A BDFRM consists of two sets of three-phase sinusoidally distributed windings with 2p and 2q poles ($p \neq q$) embedded in the same slots of stator. The rotor has no winding. However, the number of poles on rotor and stator is given by Eq. (3):

$$P_r = (p + q) \tag{3}$$

The schematic representation of BDFRM is shown in **Figure 1**. One of the two windings on stator is called as power/main winding, and the other is known as control/secondary winding. There is no direct magnetic coupling between two windings [1, 2]. The interaction between two windings takes place through rotor only. The power winding is directly excited by grid having frequency 50/60 Hz. The main role of power winding is to set up a magnetic field in the machine and deals with power exchange with grid. The control winding is excited by variable frequency, variable voltage obtained from a power electronic converter, or any other dedicated source. This winding controls torque developed and operating speed of the BDFRM. When two windings are excited, the MMFs are set up along the air gap. The MMFs F_p and F_c are the functions of rotor position where F_p and F_c are MMFS of power and control winding, respectively.

The interaction of two MMFs produces resultant air gap flux density and is responsible for development of electromagnetic torque which can be obtained analytically assuming infinite permeability of magnetic circuit, uniformly distributed windings, sinusoidal currents through windings, and representation of air gap by sine function.

The MMFs produced by these two windings are given by Eqs. (4) and (5), and the spatial distribution of power winding and control winding MMFs is shown in **Figure 2**:

$$F_p(\theta_{mg}) = F_{mp} \cos\left(\omega_p t - p\theta_{mg} + \varphi_p\right) \tag{4}$$

$$F_c(\theta_{mg}) = F_{mc} \cos\left(\omega_c t - q\theta_{mg} + \varphi_c - \alpha\right) \tag{5}$$

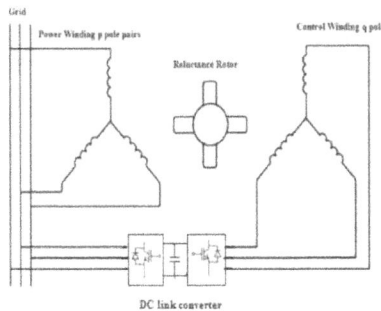

Figure 1. Schematic representation of BDFRM.

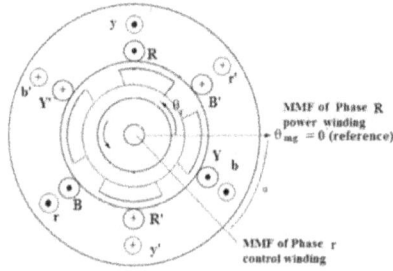

Figure 2. MMFs of power and control windings.

The resultant MMF acting along the air gap in BDFRM is the sum of MMFs produced by power and control windings. The interaction of two MMFs through rotor results in modulation of flux densities of both windings. These flux densities have fundamental frequency component along with two sidebands. Or, it can be interpreted that sinusoidal MMFs produced by two windings get modulated into a fundamental and two sidebands through inverse air gap function. The modulated frequency components are also time dependent and have spatial coordinates. Further, the coupling between two windings is decided by the pole numbers and sideband frequency component. The condition for torque production is that there is coupling between sidebands of one of the windings with fundamental frequency component of other windings. One of the sideband frequencies of power winding linking with fundamental frequency of control winding is given by Eq. (6):

$$\omega_m = (\omega_p + \omega_c)/P_r \qquad (6)$$

The space phasor voltage equations of BDFRM are used to get electromagnetic torque T_e developed which is given by Eq. (7):

$$T_e = (3/2)(p+q)\left(\psi_{dc}i_{qc} - \psi_{qc}i_{dc}\right) \qquad (7)$$

Three distinct modes of operations are possible depending upon the frequency of control winding:

1. If $\omega_c = 0$, control winding excitation is dc, and the machine operates as a conventional synchronous machine.

2. If $\omega_c > 0$, this results in super synchronous operation. The phase sequence of power winding is same as that of control winding.

3. If $\omega_c < 0$, this results in sub-synchronous operation. The phase sequence of power winding is different from that of control winding.

BDFRM is therefore potentially useful as a variable speed drive.

4. Design considerations in BDFRM

The reluctance machine has higher leakage inductances which lead to poor operational power factor. Incorporation of advanced rotors that orient flux in the preferred direction can improve the power factor [2, 14]. The other issues are smaller torque density and larger torque pulsation which can be improved by a careful design of BDFRM with higher saliency ratio. There are several considerations in design of BDFRM; the main issues are highlighted below [15, 16]:

1. What is the basis of selection of specific electric and magnetic loadings and stator slots?

2. What are the criteria for the choice of the number of poles in stator and on rotor?

3. Should there be two independent windings wound for different poles or one winding wound for different poles in stator?

4. Do the stator windings have to be full pitched or short pitched?

5. Which of the two windings should be selected as power winding/control winding?

6. What will be the practical range of control winding frequency and control winding voltage?

7. Should the choice of air gap length based on magnetic pull or any other factor?

8. Is there any possibility of reduction of space harmonics, noise, losses in stator windings, and rotor core?

9. How to reduce or mitigate bearing currents due to PWM inverters and unbalanced magnetic pull?

The selection of specific loadings depends on permissible losses, overload capacity, magnetic saturation, forces acting on rotor surface, etc. Selection of the number of stator slots and rotor ducts is very important in elimination of space harmonics in resultant MMF waveform. As the resultant MMF wave along the air gap is superposition of two MMFs, it has large space harmonics. **Table 3** gives suitable combinations of stator slots and rotor ducts.

Proper selection of the number of poles on stator and rotor decides the magnetic coupling between stator and rotor which governs the torque production. The higher the coupling coefficient, the higher the torque production. This can be inferred from **Figure 3**.

S. no.	Stator slots	Rotor ducts	Lower-order space harmonics	
			h_{sp}	h_{sq}
1	48	54	104	104
2	48	60	52	56
3	48	66	136	140
4	48	72	136	140

Table 3. Minimum number of space harmonics for same stator and rotor slots [17, 18].

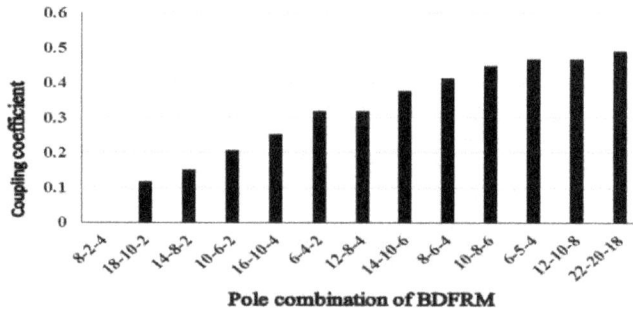

Figure 3. Coupling coefficient vs. pole combinations of BDFRM.

The selection of two or single winding in the stator is done on the basis of relative merits of both cases. Though it is costly, provision of two independent windings is more convenient for construction and control. With two independent windings, there is a great flexibility in winding pitches, and short pitching can be used to minimize the effect of space harmonics. The higher pole winding is generally considered as power winding [15] though there is no definite rule. There is no clarity over the voltage rating of control winding. It can be understood that in V/f control mode voltage will vary as frequency to avoid saturation in magnetic circuit. Is V_c/f_c equal to V_p/f_p? Or, can it be different? The length of air gap greatly affects the performance of the machine as MMF requirement increases with the length of air gap. Hence, selection of air gap needs a compromise.

Design of BDFRM is a complex process and involves various aspects of electromagnetism, thermal engineering, and machine design. The thermal and mechanical engineering aspects are similar to conventional machine design criteria. In BDFRM mechanical geometries differ from conventional machines, and hence conventional design procedures may have to be modified suitably. Therefore, analytical design approach is used. The voltage and current ratings of BDFRM decide the capacity of power electronic converter. The major consideration is the cost of machine which depends on the volume of active material used. For continuous operation the torque should be present over the designed range of speed, and the losses in BDFRM should be kept minimum for restricting temperature rise.

4.1. Steps in the design of BDFRM

For the design of BDFRM rating of the machine (2 kW), a number of pole configurations on stator and rotor (8-6-4), rated synchronous speed (500 rpm), rotor construction (ducted), etc. are specified. With these inputs the design of the machine can be carried out on the guidelines suggested in [15, 16, 19]. After deciding the rating and the input data of BDFRM, the steps involved in the design are shown as a flow graph in **Figure 4**. Two configurations of BDFRM are short listed for prototype development: 6-4-2 with reluctance rotor, and the other one is 8-6-4 circular ducted rotor. Initial values of flux density (0.5 T), current density (4 A/mm²), integral slot, and full-pitched windings are assumed. A nonlinear programing (NLP) method is selected for design optimization of BDFRM as nonlinearities of materials and magnetic circuit have to be taken into considerations.

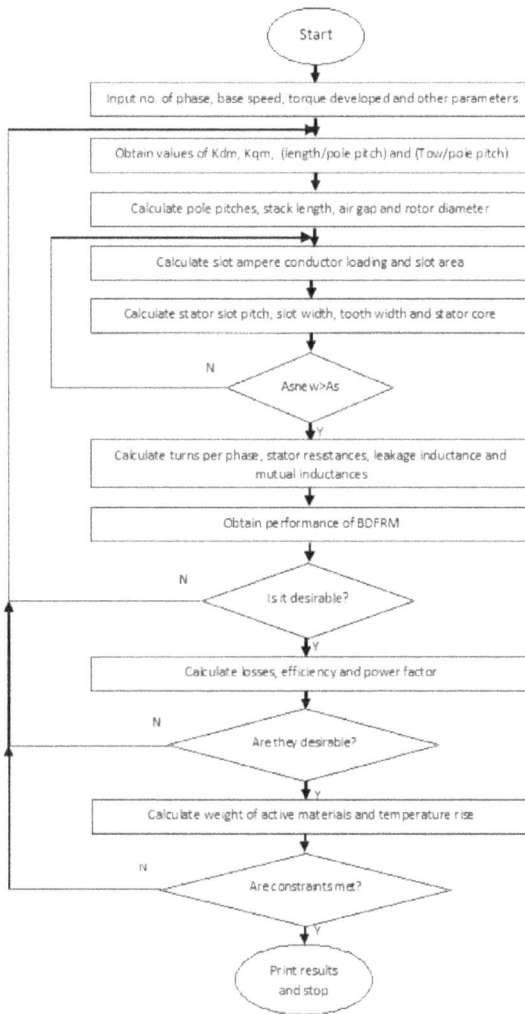

Figure 4. Flow graph for design optimization.

4.2. Objective function

A NLP problem involves the selection of independent variables, the development of objective function, and the constraint function for design of electrical machine. The objective function can be the cost of active materials or overall weight of machine or performance parameters, etc. The performance parameters of a machine can be used as constraints, e.g., pullout torque/full-load power factor/slot space factor/specific loadings/maximum flux density in stator tooth/temperature rise/shaft diameter. While designing BDFRM only a few variables are considered as independent, and these are given below [6, 19]:

1. Stator bore radius (x_1)

2. Gross length of stator stack (x_2)

3. Width of stator slot (x_3)

4. Depth of stator slot (x_4)

5. Cross section of winding conductors (x_5)

6. Depth of stator core (x_6)

7. Minimum length of air gap (x_7)

8. Maximum length of air gap (x_8)

9. Number of turns per pole per phase (x_9)

10. Number of slots in stator (x_{10})

The variables such as height of tooth lip, slot wedge thickness, slot opening, etc. have little effect on the overall performance. They are assumed to be known and constant and hence not considered in the optimization. The variables considered in BDFRM design are shown in **Figure 5**.

The constraint function limits considered in the design of BDFRM are:

1. Maximum torque ≥ 1.5 time full-load torque

2. Full-load power factor ≥ 0.5 lagging

3. Current density ≤ 4.5 A/mm^2

4. Slot fill factor ≤ 65%

5. Maximum tooth flux density ≤ 2 T

6. Maximum flux density in stator core and yoke ≤ 1.3 T

7. Specific electrical loading ≤ 25,000 ampere-conductor/m

8. Width of stator slot ≥ 6 mm

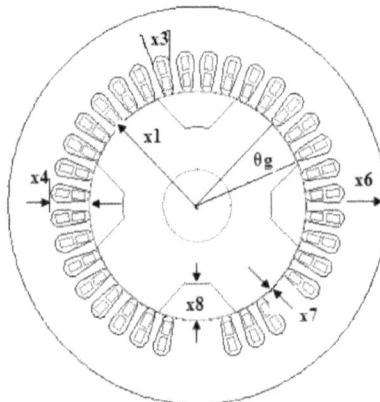

Figure 5. Illustration of the design variables of BDFRM.

9. Temperature rise ≤ 50°C

10. Shaft diameter ≥ 25 mm

The objective function [20] for reluctance rotor configuration is derived as given below.

The outer radius of stator (S_r) lamination is given by Eq. (8):

$$S_r = (x_1 + x_4 + x_6)$$ (8)

The volume of stator laminations is calculated by Eq. (9):

$$V_{is} = L_i \left[\pi(x_1 + x_4 + x_6)^2 - \pi x_1^2 - x_3\, x_4 x_{10} \right]$$ (9)

where $L_i = 0.9(x_2 - n_d b_d)$.

Thereafter, volume of rotor is calculated by using Eq. (10):

$$V_{ir} = 0.9x_2 \left(\pi x_1^2 - 4x_1^2 \left(\theta_g - sin\theta_g\, cos\theta_g \right) \right)$$ (10)

Approximate mean length of turns for power and control windings are given by Eqs. (11) and (12), respectively:

$$L_{mtp} = 2x_2 + 2.3\pi(x_1/p) + 0.152$$ (11)

$$L_{mtc} = 2x_2 + 2.3\pi(x_1/q) + 0.152$$ (12)

The volume of the copper required for power and control windings is given by Eq. (13):

$$V_c = 2\left(pmx5.x9.L_{mtp} + qmx_5 x_9 L_{mtc} \right)$$ (13)

Finally, the objective function is obtained as in Eq. (14) taking into account of the weights of active materials:

$$F = D_i(V_{is} + V_{ir}) + D_c V_c$$ (14)

For ducted rotor BDFRM, rotor iron volume is derived as given by Eq. (15):

$$V_{ir} = 0.9x_2 \left(\pi x_1^2 - \pi\, d_{sh}^2 \right) - 6(A_1 + A_2 + A_3 + A_4 + A_5 + A_6)$$ (15)

where A_1–A_6 represent duct areas and are calculated on the basis of width of the duct and angular distribution over a rotor pole.

Eq. (15) is substituted in place of V_{ir} in Eq. (14) to get the objective function for ducted rotor BDFRM.

The specifications of two 2 kW BDFRM configurations are given below:

a. Power winding poles—6, control winding poles—2, rotor poles(reluctance)—4, rated speed 750 rpm, stator slots 36 with single winding in stator

Sr.	Description	Initial values	Optimized 6-4-2 BDFRM
1	Stator outer diameter	204 mm	165 mm
2	Stator internal diameter (stator bore)	113.29 mm	91 mm
3	Length of air gap (main)	0.45 mm	0.5 mm
4	Rotor external diameter	112 mm	90 mm
5	Effective axial length of machine	95	80 mm
6	Stator yoke length	17 mm	15.3 mm
7	Rotor inner diameter	40 mm	50 mm
8	Conductor area for both windings	0.665 mm^2	0.663 mm^2
9	Depth of stator slot	21.9 mm	18.016 mm
10	Stator tooth width	3 mm	4 mm
11	Slot area	79.79 mm^2	69.25 mm^2
12	Total slot space factor	0.45	0.45
13	(ac) Loading of power/control winding	5000 A/m	5140 A/m
14	Max flux density in stator tooth	1.46 T	1.3 T
15	Max flux density in stator yoke	1.2 T	1.3 T
16	Max flux density in rotor tooth	1.2 T	1.45 T
17	Gap flux density	0.5 T	0.73 T
18	Weight of copper	3.564 kg	2.494 kg
19	Weight of stator core	4.84 kg	2.262 kg
20	Weight of stator teeth	2.84 kg	2.123 kg
21	Weight of rotor	2.8 kg	1.5 kg

Table 4. Particulars of 6-4-2 configuration of BDFRM.

Sr.	Description	Initial values	Optimized 8-6-4 BDFRM
1	Stator outer diameter	235 mm	210 mm
2	Stator internal diameter (stator bore)	169 mm	144.24 mm
3	Length of air gap (main)	0.45 mm	0.5 mm
4	Rotor external diameter	168 mm	143 mm
5	Effective axial length of machine	106	69 mm
6	Stator yoke length	9 mm	14.5 mm
7	Rotor pole pitch	85.37 mm	75.5 mm
8	Rotor inner diameter	30 mm	35 mm
9	Area of power winding conductor	0.506 mm^2	0.653 sq.mm dia. 0.9118 mm
10	Area of control winding conductor	0.506 mm^2	0.653 sq.mm dia. 0.9118 mm
11	Depth of stator slot	18 mm	26.88 mm
12	Stator tooth width	5 mm	5.1278 mm

Sr.	Description	Initial values	Optimized 8-6-4 BDFRM
13	Slot area	113 mm^2	177.32 sq. mm
14	Total slot space factor	0.45	0.35
15	(ac) Loading of power winding	10 kA/m	12.4 kA/m
16	(ac) Loading of control winding	10 kA/m	9.04 kA/m
17	Max flux density in stator tooth	1.2 T	1.48 T
18	Max flux density in stator yoke	1.2 T	1.48 T
19	Max flux density in rotor tooth	1.2 T	1.38 T
20	Gap flux density	0.5 T	0.5 T
21	Weight of copper	4.559 kg	3.783 kg
22	Weight of stator core	4.989 kg	2.99 kg
23	Weight of stator teeth	3.911 kg	2.86 kg
24	Weight of rotor laminations	7.85 kg	6.8 kg

Table 5. Particulars of 8-6-4 configuration of BDFRM.

b. Power winding poles—8, control winding poles—4, rotor poles(ducted)—6, rated speed 500 rpm, stator slots 48 with two independent windings in stator

A nonlinear optimization technique is based on constrained optimization along with finite element analysis using Maxwell 15 2D software for design optimization of BDFRM. The key dimensions of BDFRM prototypes which are obtained after design optimization are given in **Tables 4** and **5**.

In optimized prototypes of 6-4-2 and 8-6-4 BDFRMs, requirements of active materials have reduced by 50 and 29% respectively. The considerable reduction in active material for 6-4-2 BDFRM is due to the use of single winding in stator. Even though material requirements have gone down, it hardly affects the performance parameters. This can be observed from **Figures 6–8** where the plots for power factor, efficiency, and torque vs. weight of active materials used in

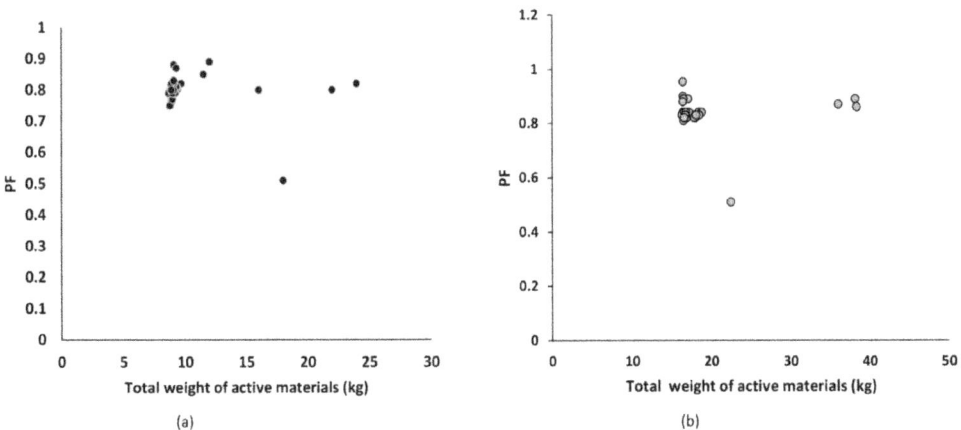

(a)

(b)

Figure 6. Variation of power factor with weight of active materials. (a) 6-4-2 BDFRM and (b) 8-6-4 BDFRM.

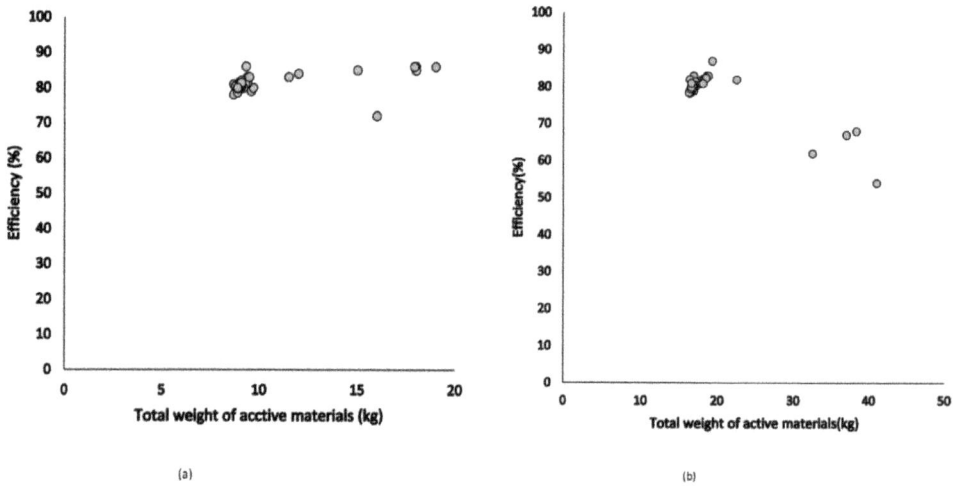

Figure 7. Variation of efficiency with weight of active materials. (a) 6-4-2 BDFRM and (b) 8-6-4 BDFRM.

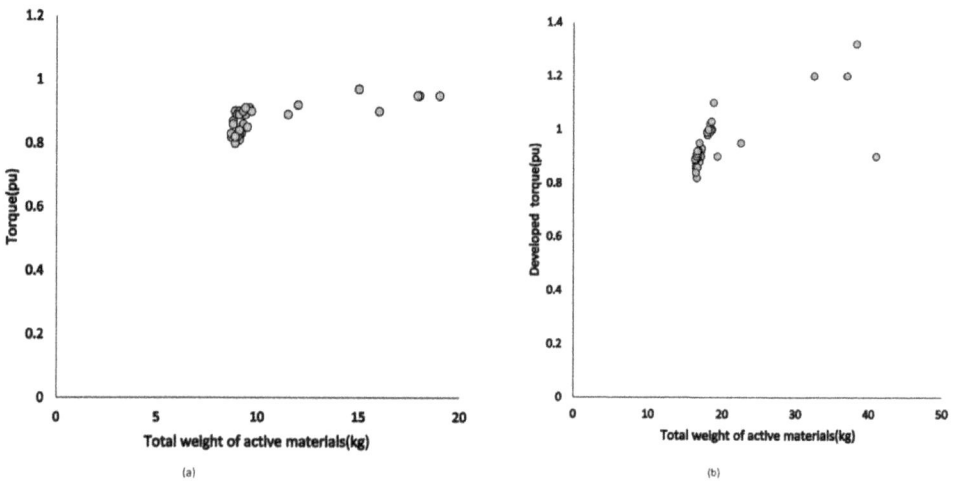

Figure 8. Variation of developed torque (pu) with weight of active materials. (a) 6-4-2 BDFRM and (b) 8-6-4 BDFRM.

BDFRM for all candidate machine designs were evaluated during optimization process, respectively. The data points for these variables form a cluster around the optimum value of active materials.

5. Prototypes of BDFRM

Two prototypes of BDFRM are fabricated based on optimum design, and details of which are given below [21].

5.1. 6-4-2 Reluctance rotor configuration

The stator has 36 slots in which a single winding is embedded. The details of stator and rotor laminations (thickness 0.35 mm) used are shown in **Figure 9**.

The details of winding are shown in **Figure 10**. The winding is developed such that two different poles (6 and 2) are formed with single winding. The end connections and coil groups

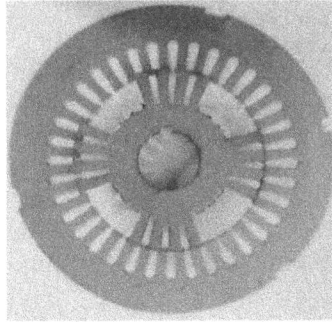

Figure 9. Laminations used in 6-4-2 BDFRM.

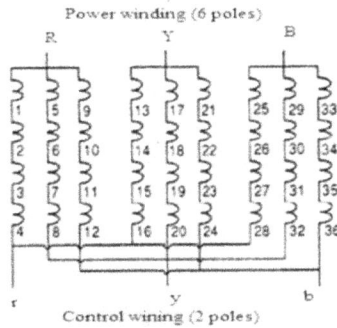

Figure 10. Winding arrangement in 6-4-2 BDFRM.

Figure 11. Fabrication details of 6-4-2 BDFRM.

are connected such that one end behaves as a 6-pole (power) winding and the other end acts as a 2-pole (control) winding. A short-pitched (short pitched by one slot) double-layer winding with 36 coils is designed primarily for two poles. Nine groups are formed with each group having four coils. The start and end terminals of each coil group are brought out for giving flexibility in winding connection. This has resulted in increased length of end connections than the normal case. Hence, special size end covers have to be designed. The actual photographs of 6-4-2 BDFRM during fabrication are shown in **Figure 11**.

5.2. 8-6-4 Circular ducted rotor configuration

8-6-4 Circular ducted rotor BDFRM is fabricated. Stator consists of two independent windings one designed for eight poles (power) and the other for four poles (control). The windings are star connected. The stator windings are accommodated in 48 slots to get full-pitched windings. The lamination has deeper slots so that two double-layer windings can be fitted with a good slot space factor. The use of two double-layer windings facilitated the arrangement of end connections. All terminals of the windings are brought out so as to get flexibility in connection.

The number of ducts per rotor pole is selected as 10 as per the guidelines given in **Table 3** [16, 17]. The fabrication details are shown in **Figure 12**.

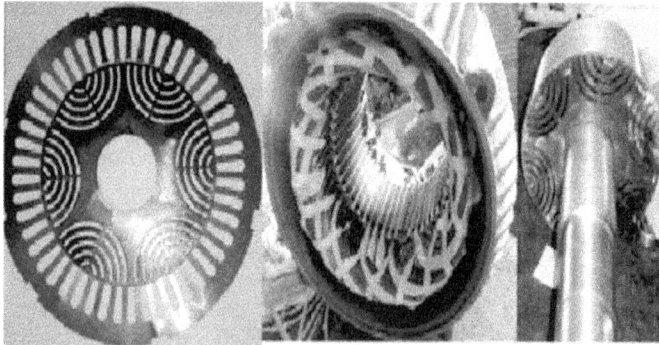

Figure 12. Fabrication details of 8-6-4 BDFRM.

6. Finite element analysis of BDFRM

By using MAXWEL 16 software, the finite element models [22] are developed from the actual dimensions of stator and rotor laminations for 6-4-2 and 8-6-4 configurations. The models are simulated to get flux density distribution, torque, and surfaces forces. The flux density distribution in 6-4-2 BDFRM and 8-6-4 BDFRM is shown in **Figure 13(a)** and **(b)**, respectively. It may be observed that flux density values are staying within saturation limit. The peak magnitude of flux density in prototype BDFRMs does not exceed 1.79 T.

Reluctance machines develop weak and pulsating torque. BDFRM is not different. This can be seen from **Figure 14(a)**. However, the torque developed by ducted rotor BDFRM is higher with reduced pulsation as shown in **Figure 14(b)**.

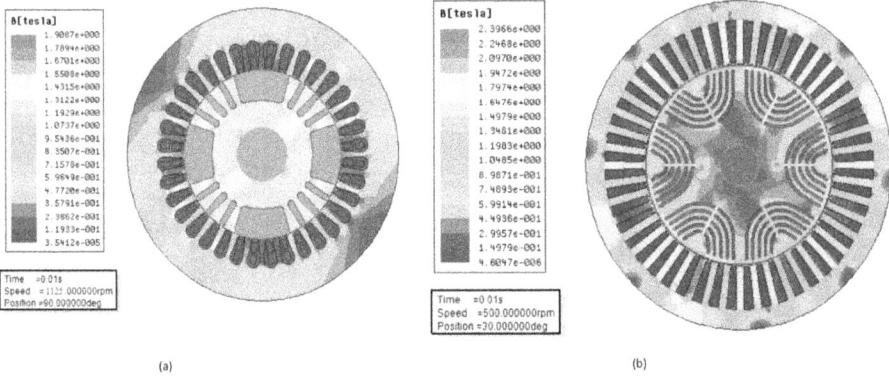

Figure 13. Flux density distribution in BDFRM. (a) 6-4-2 BDFRM and (b) 8-6-4 BDFRM.

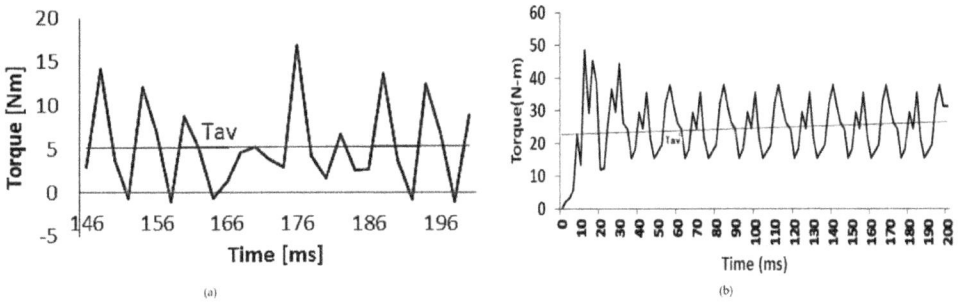

Figure 14. Torque developed by BDFRM. (a) 6-4-2 BDFRM and (b) 8-6-4 BDFRM.

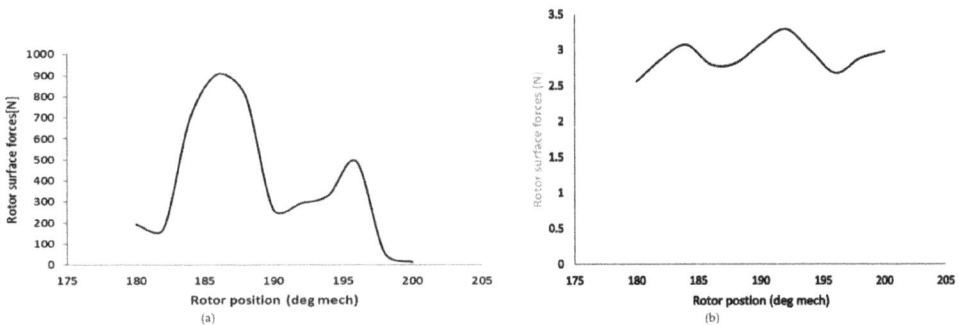

Figure 15. Unbalanced magnetic pull acting on rotor surface. (a) 6-4-2 BDFRM and (b) 8-6-4 BDFRM.

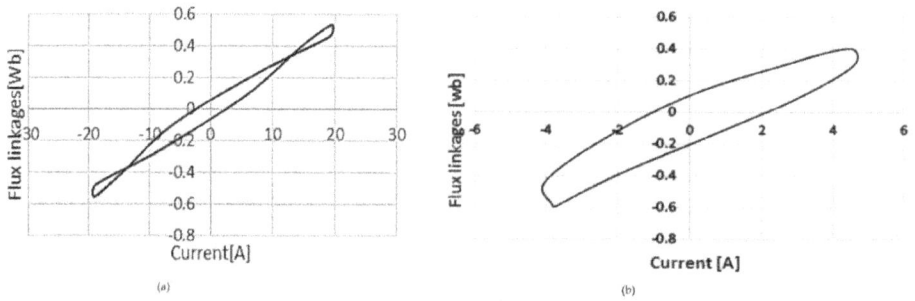

Figure 16. λ-I plots for BDFRM. (a) 6-4-2 BDFRM and (b) 8-6-4 BDFRM.

Due to the absence of winding on rotor in BDFRM, there is no counterbalancing of MMF for stator MMF. This develops unbalanced magnetic forces on rotor surface. The magnitude of forces is quite large in reluctance rotor configuration, whereas they are considerably reduced in case of ducted rotor. This can be seen from **Figure 15**. The magnetic forces greatly reduce with advanced rotor configurations.

The flux linkage-current plots (λ-i) for both configurations are shown in **Figure 16**. The performance of machine greatly depends on area of (λ-i) plot. The plots clearly indicate that 8-6-4 BDFRM has better performance due to larger area of (λ-i) plot.

7. Performance of prototypes

The performances of two prototype BDFRMs are obtained from load tests. BDFRM is coupled with a dc machine of rating 2 kW, 220 V, and 1500 rpm which acts as a load. A three-phase two-level, 3 kW, IGBT variable voltage variable frequency inverter is used for exciting the control

6-4-2 Reluctance rotor BDFRM		Description	8-6-4 Ducted rotor BDFRM	
Optimized machine (calculated)	Actual value@60% loading		Optimized machine (calculated)	Actual value@60% loading
2 kW	1.2 kW	Rated output	2 kW	1.6 kW
750 rpm	750 rpm	Base speed	500 rpm	500 rpm
80%	75%	Efficiency	83%	75%
0.7	0.54	Power factor	0.8	0.7
25.46 N-m	15.27 N-m (from FEA analysis)	Torque developed	38 N	30 N (from FEA analysis)
1.35 T	1.77 T (from FEA analysis)	Flux density in stator tooth	1.48 T	1.77 T (from FEA analysis)
34°C	42°C (by resistance method)	Temperature rise	34.45°C	45°C (by resistance method)

Table 6. Performance details of prototype BDFRMs.

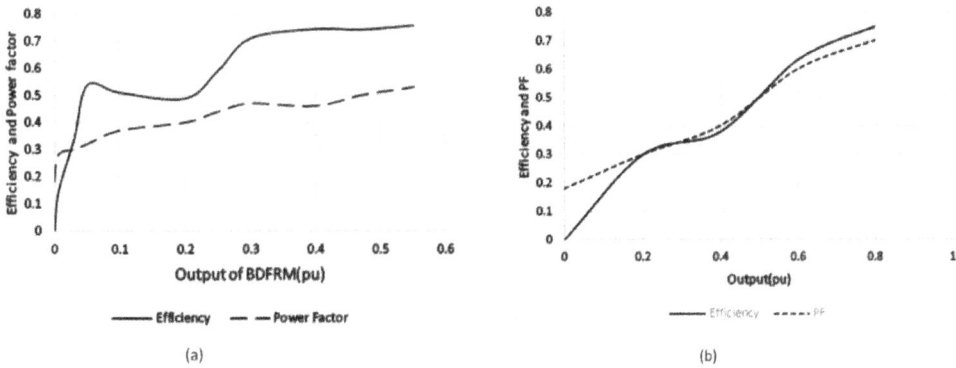

Figure 17. Performance curves of prototype BDFRMS. (a) 6-4-2 BDFRM and (b) 8-6-4 BDFRM.

winding. The results are presented in **Table 6**. Because of the limitation of laboratory test facilities, loading could be done up to 70%.

The capabilities of prototype BDFRMs can be judged from the performance curves shown in **Figure 17**. There is slight deviation of parameters from designed values. This may be due to practical difficulties during loading the machine. Additionally, there is an increase in losses due to the presence of harmonics in control winding excitation which has compromised efficiency and power factor. This advocates development of sophisticated control algorithm for BDFRM to get the desired performance.

8. Conclusion

This chapter highlights various optimization methods, suitability of nonlinear programming methods, and recent stochastic or population-based methods for optimal design of electrical machines. It also discusses briefly a few issues in the design of BDFRM. An objective function is developed for minimization of active material requirement, and an algorithm based on nonlinear programming is presented for optimal design of 2 kW BDFRMs. The analytical models developed based on optimized design are simulated in Maxwell 16 software. The simulation results closely agree with the design values, and there is no saturation in magnetic circuit. The field tests on the two prototypes have demonstrated the capabilities of BDFRMs. It is observed that the performance of 8-6-4 BDFRM is better than 6-4-2 BDFRM in all respects. Both the machines have attained rated speed. Speed control has been achieved on either side of it. Although efficiencies of the prototypes are closer to design values, power factor is lower than expected.

Author details

Mandar Bhawalkar*, Gopalakrishnan Narayan and Yogesh Nerkar

*Address all correspondence to: mandar_bhawalkar@yahoo.co.in

PVG's College of Engineering and Technology, Pune, India

References

[1] Betz R, Jovanovic M. Theoretical analysis of control properties for brushless doubly fed reluctance machine. IEEE Transactions on Energy Conversion. Sept 2002;**17**(1):332-339. DOI: 10.1109/TEC.2002.801997

[2] Betz R, Jovanovic M. The brushless doubly fed reluctance machine and the synchronous reluctance machine—a comparison. IEEE Transactions on Industry Applications. Jul/Aug 2000;**36**(4):1103-1110. DOI: 10.1109/28.855966

[3] Stipetic S, Miebach W, Zarko D. Optimization in design of electrical machines: Methodology and workflow. In: Proceedings of the International Conference on ACEMP-OPTIM-Electromotion Conference; 2015. pp. 441-448

[4] Liu A, Xu W. A global optimization approach for electrical machine designs. IEEE Power & Energy Society General Meeting. 2007:1-8. ISBN: 1-4244-1298-6

[5] Rao S. Engineering Optimization Theory and Practice. 3rd ed. New Delhi: New Age International Publisher; 2013. 722p. ISBN 978-81-224-2723-3

[6] Ramamoorthy M. Computer Aided Design of Electrical Equipments. Reprint. New Delhi: Affiliated East West Press Pvt Ltd; 2011, 2011. pp. 5-53. ISBN 81-85095-57-4

[7] Ma C, Qu L. Multiobjective optimization of switched reluctance motors based on design of experiments and particle swarm optimization. IEEE Transactions on Energy Conversion. Sept. 2015;**30**(3):1144-1153. DOI: 10.1109/TEC.2018.2411677

[8] Jiang W, Jahns T, Lipo T, Taylor W, Suzuki Y. Machine design optimization based on finite element analysis in a high-throughput computing environment. In: Proceedings of IEEE Energy Conversion Congress and Exposition; Sept. 2012. 10.1109/ECCE.2012.6342727

[9] Legranger J, Friedrich G, Vivier S, Mipo J. Combination of finite-element and analytical models in the optimal multidomain design of machines: Application to an interior permanent magnet starter generator. IEEE Transactions on Industry Applications. Jan 2010;**46**(1): 232-239. DOI: 10.1109/TIA.2009.2036549

[10] Ponmurugan P, Rengarajan N. Multiobjective optimization of electrical machines, a state of the art study. Journal of Computer Applications. Oct. 2012;**56**(13):26-30. DOI: 10.1.1.244.6959

[11] Idir K, Chang L, Dai H. A neural network based optimization approach for induction motor design. Canadian Conference on Electrical & Computer Engineering; May1996. pp. 951-954

[12] Bétin F, Yazidi A, Sivert A, Fuzzy CG. DOI n DOI logic control design for electrical machines. International Journal of Electrical Engineering and Technology. May–June 2016; 7:14-24. ISSN 0976-6545

[13] Çuncaş M. Design optimization of electric motors by multiobjective fuzzy genetic algorithm. Mathematical and Computational Applications. 2008;**13**(3):153-163

[14] Liao Y, Xu L, Li Z. Design of a doubly fed reluctance motor for adjustable speed drive. IEEE Transactions on Industry Applications. Sept/Oct 1996;**32**(5):1195-1203. DOI: 10.1109/28.536883

[15] Knight A, Betz R, Dorrell D. Issues with the design of brushless doubly fed reluctance machines: Unbalanced magnetic pull, skew and iron losses. 2011 IEEE International Electric Machines & Drives Conference (IEMDC); 2011. pp. 663-668

[16] Knight A, Betz R, Dorrell D. Design and analysis of brushless doubly fed reluctance machines. IEEE Transactions on Industry Applications. Jan/Feb 2013;**49**:50-57. DOI: 10.1109/TIA.2012.2229451

[17] Vagati A, Franceschini G, Marongiu I, Troglia G. Design criteria of high performance synchronous reluctance motors. IAS Annual Meeting. 1992;**1**:66-73

[18] Vagati A, Pastorelli M, Francheschini G, Petrache S. Deign of low torque ripple synchronous reluctance motor. IEEE Transactions on Industry Applications. 1998;**34**(4):758-765. DOI: 10.1109/IAS.1997.643040

[19] Boldea I. Reluctance Synchronous Machines and Drives. New York, USA: Oxford Science Publications; 1996. ISBN: 0 19 85391 0

[20] Kunte S, Bhawalkar M, Gopalakrishnan N, Nerkar Y. Optimal design and comparative analysis of different configurations of brushless doubly fed reluctance machine. IEEJ Transactions on Industry Applications. Nov. 2017;**6**(6):370-380. DOI: 10.1541/ieejjia.6.370

[21] Bhawalkar M. Studies in Wind Power Generation Systems [Ph.D. Thesis]. India: Savitribai Phule Pune Univeristy; Oct. 2017

[22] Bianchi N. Electrical Machine Analysis Using Finite Elements. FL, USA: Taylor and Francis, Special Indian reprints; 2015. ISBN 9780849333996

Control of Induction Machines

Zero and Low-Speed Sensorless Control of Induction Machines Using Only Fundamental Pulse Width Modulation Waveform Excitation

Qiang Gao, Greg Asher and Mark Sumner

Additional information is available at the end of the chapter

http://dx.doi.org/10.5772/intechopen.75892

Abstract

This chapter presents a position sensorless method for induction machines that only relies on the fundamental pulse width modulation (PWM) waveforms to excite saliency. Position signals can be synthesized through the measurement of the derivatives of the line currents induced by the PWM voltage vectors. This method is essentially saliency detection based, and therefore derivation of the rotor position is possible at low and zero speeds. In addition, it works also at higher speeds without the need of the knowledge of the machine's fundamental model. Experimental results showing fully sensorless induction motor control at low and higher speeds validate the principle of this method.

Keywords: sensorless control, induction machine, saliency, PWM excitation, current derivative

1. Introduction

Due to the incapacity of the fundamental model-based sensorless rotor position estimation at zero and low frequencies, alternative methods have been intensively studied. These methods exploit the anisotropy or saliency of the machine resulting from either saturation or geometric variation on the rotor. They can be classified into two categories according to the detection method for the anisotropy (or saliency) position. One category relies on the continuous injection of high-frequency voltage signals and then measuring the response of the high-frequency (*hf*) current [1–7]. Demodulation of the *hf* current signal enables the extraction of the rotor angle. The second category makes use of the line-current transient response to a PWM switching state. This can be realized by injecting special voltage test vectors [8–13] or by modifying the normal pulse width modulation waveforms

IntechOpen

[14], which may increase the hazardous common mode current in the machine [15]. The induced current transient response during a test vector reflects the spatial variation of the stator leakage inductances due to the anisotropy. Therefore, it is possible to acquire the rotor position, or rotor flux angle, through the measurement of the transient current derivative in response to the test vector.

In this chapter, a method belonging to the second category is described. Instead of using extra test vectors or modifying the standard modulation scheme, this method integrates the test vectors with the standard PWM waveforms [16]. In the following paragraphs, the theory of the method will be presented first, then, its application on a 4-pole 30 kW Δ-connected cage machine having 56 open slots will be demonstrated. Other implementation issues related to the speed sensorless operation, such as the noise filter of the position signals, will also be introduced.

2. Position estimation with the fundamental wave PWM

When a three-phase, delta-connected induction machine has its stator leakage inductances modulated by the anisotropies introduced by either the main flux saturation or the rotor slotting, they can be assumed to vary according to:

$$l_{\sigma a} = l_0 + \Delta l \cos(n_{an}\theta_{an}) \tag{1}$$

$$l_{\sigma b} = l_0 + \Delta l \cos(n_{an}(\theta_{an} - 2\pi/3)) \tag{2}$$

$$l_{\sigma c} = l_0 + \Delta l \cos(n_{an}(\theta_{an} - 4\pi/3)) \tag{3}$$

where l_0 is the average inductance and Δl is the amplitude of inductance variation caused by the anisotropy ($n_{an} = 2$ for saturation-induced anisotropy or $n_{an} = n_{rs} = N_r/p$ for rotor slotting, where N_r is rotor slot number and p the pole pairs).

The standard space vectors of **Figure 1** are applied. **Figure 2** shows the equivalent circuit when the machine is applied with vector $\underline{u}_1$ from which the following equations can be derived:

$$U_d = i_{ab}^{(u1)}r_s + l_{\sigma a}\frac{di_{ab}^{(u1)}}{dt} + e_a^{(u1)} \tag{4}$$

$$0 = i_{bc}^{(u1)}r_s + l_{\sigma b}\frac{di_{bc}^{(u1)}}{dt} + e_b^{(u1)} \tag{5}$$

$$-U_d = i_{ca}^{(u1)}r_s + l_{\sigma c}\frac{di_{ca}^{(u1)}}{dt} + e_c^{(u1)} \tag{6}$$

By the application of the null vector $\underline{u}_0$ or $\underline{u}_7$, one has:

$$0 = i_{ab}^{(u0)}r_s + l_{\sigma a}\frac{di_{ab}^{(u0)}}{dt} + e_a^{(u0)} \tag{7}$$

$$0 = i_{bc}^{(u0)}r_s + l_{\sigma b}\frac{di_{bc}^{(u0)}}{dt} + e_b^{(u0)} \tag{8}$$

$\underline{u}_3(0,1,0)$

$\underline{u}_2(1,1,0)$

II

III

I

$\underline{u}^*$

$\underline{u}_4(0,1,1)$

$\underline{u}_1(1,0,0)$

IV

VI $\underline{u}_0(0,0,0)$

$\underline{u}_7(1,1,1)$

V

$\underline{u}_5(0,0,1)$

$\underline{u}_6(1,0,1)$

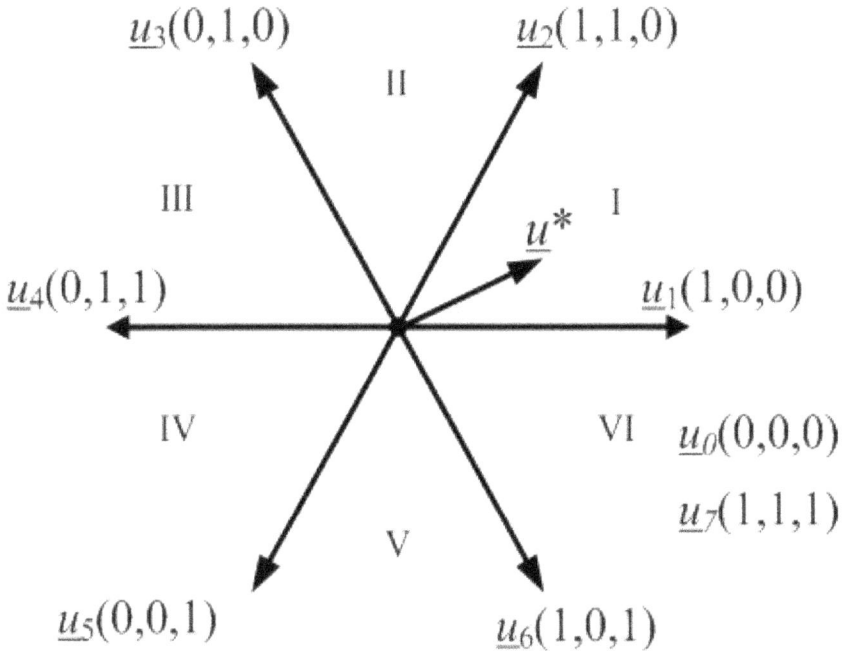

Figure 1. Definition of space vectors.

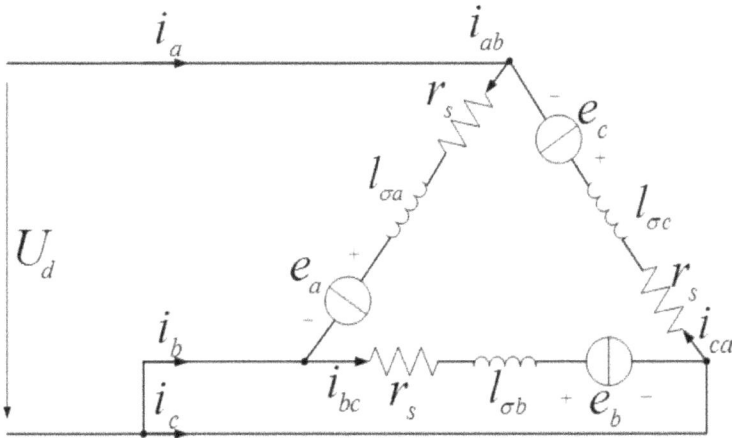

i_a

i_{ab}

r_s

e_c

$l_{\sigma a}$

$l_{\sigma c}$

U_d

e_a

r_s

i_b

i_{ca}

i_c

i_{bc} r_s

$l_{\sigma b}$ $+$ e_b $-$

Figure 2. Equivalent circuit with $\underline{u}1$ being applied.

$$0 = i_{ca}^{(u0)} r_s + l_{\sigma c} \frac{d i_{ca}^{(u0)}}{dt} + e_c^{(u0)} \qquad (9)$$

If the instants of applying $\underline{u}_1$ and $\underline{u}_0$ are close enough, it is viable to assume that:

$$e_a^{(u0)} \approx e_a^{(u1)}, e_b^{(u0)} \approx e_b^{(u1)}, e_c^{(u0)} \approx e_c^{(u1)}.$$

Additionally, the voltage drops across the stator resistance and can be ignored due to their small values compared with U_d.

Hence, the subtraction of the Eqs. (4)–(7), (5)–(8) and (6)–(9) yields:

$$\frac{di_{ab}^{(u1)}}{dt} - \frac{di_{ab}^{(u0)}}{dt} = \frac{U_d}{l_{\sigma a}} \tag{10}$$

$$\frac{di_{bc}^{(u1)}}{dt} - \frac{di_{bc}^{(u0)}}{dt} = 0 \tag{11}$$

$$\frac{di_{ca}^{(u1)}}{dt} - \frac{di_{ca}^{(u0)}}{dt} = -\frac{U_d}{l_{\sigma a}} \tag{12}$$

From the relationship between phase currents and line currents, for example, $i_a = i_{ab} - i_{ca}$, one has:

$$\frac{di_a^{(u1)}}{dt} - \frac{di_a^{(u0)}}{dt} = \frac{l_{\sigma a} + l_{\sigma c}}{l_{\sigma a} l_{\sigma c}} U_d \tag{13}$$

$$\frac{di_b^{(u1)}}{dt} - \frac{di_b^{(u0)}}{dt} = -\frac{1}{l_{\sigma a}} U_d \tag{14}$$

$$\frac{di_c^{(u1)}}{dt} - \frac{di_c^{(u0)}}{dt} = -\frac{1}{l_{\sigma c}} U_d \tag{15}$$

Considering Eqs. (1) to (3), one has:

$$\frac{di_a^{(u1)}}{dt} - \frac{di_a^{(u0)}}{dt} = \frac{U_d}{l_0}\left(2 + \frac{\Delta l}{l_0}\cos\left(n_{an}\left(\theta_{an} - \frac{2\pi}{3}\right)\right)\right) \tag{16}$$

$$\frac{di_b^{(u1)}}{dt} - \frac{di_b^{(u0)}}{dt} = -\frac{U_d}{l_0}\left(1 - \frac{\Delta l}{l_0}\cos(n_{an}\theta_{an})\right) \tag{17}$$

$$\frac{di_c^{(u1)}}{dt} - \frac{di_c^{(u0)}}{dt} = -\frac{U_d}{l_0}\left(1 - \frac{\Delta l}{l_0}\cos\left(n_{an}\left(\theta_{an} - \frac{4\pi}{3}\right)\right)\right) \tag{18}$$

from which three balanced position scalars p_a, p_b and p_c can be defined as follows:

$$p_a = 1 + c_1\left(\frac{di_b^{(u1)}}{dt} - \frac{di_b^{(u0)}}{dt}\right) \tag{19}$$

$$p_b = -2 + c_1\left(\frac{di_a^{(u1)}}{dt} - \frac{di_a^{(u0)}}{dt}\right) \tag{20}$$

$$P_c = 1 + c_1 \left(\frac{di_c^{(u1)}}{dt} - \frac{di_c^{(u0)}}{dt} \right) \tag{21}$$

where $c_1 = l_0/U_d$. If both c_1 and $\frac{di}{dt}$ are known, it is possible to construct the position vector p directly via:

$$\underline{p} = p_a + a \cdot p_b + a^2 \cdot p_c \tag{22}$$

where $a = e^{j2\pi/3}$. However, c_1 consists of an unknown coefficient l_0, which may vary with the saturation level of the main flux. This uncertainty can be avoided by looking at the current response to another voltage vector $\underline{u}_2$. Following the same as earlier, another three position scalars can be defined as:

$$P_a = -2 - c_1 \left(\frac{di_c^{(u2)}}{dt} - \frac{di_c^{(u0)}}{dt} \right) \tag{23}$$

$$P_b = 1 - c_1 \left(\frac{di_b^{(u2)}}{dt} - \frac{di_b^{(u0)}}{dt} \right) \tag{24}$$

$$P_c = 1 - c_1 \left(\frac{di_a^{(u2)}}{dt} - \frac{di_a^{(u0)}}{dt} \right) \tag{25}$$

It should be noted that because the application of $\underline{u}_1$ and $\underline{u}_2$ results in the same position scalars, they are defined with the same terms, p_a, p_b and p_c.

By referring to (19), (20), (21) and (23), (24), (25), it is possible to define $\underline{p}$ through the combination of (19), (24) and (25):

$$
\begin{aligned}
\underline{p} = p_\alpha + jp_\beta &= p_a + ap_b + a^2 p_c \\
&= c_1 \left[
\begin{array}{l}
\left(\dfrac{di_b^{(u1)}}{dt} - \dfrac{di_b^{(u0)}}{dt} \right) - a \left(\dfrac{di_b^{(u2)}}{dt} - \dfrac{di_b^{(u0)}}{dt} \right) \\[2mm]
- a^2 \left(\dfrac{di_a^{(u2)}}{dt} - \dfrac{di_a^{(u0)}}{dt} \right)
\end{array}
\right]
\end{aligned}
\tag{26}
$$

Therefore,

$$
p_\alpha = c_1 \left[
\begin{array}{l}
\left(\dfrac{di_b^{(u1)}}{dt} - \dfrac{di_b^{(u0)}}{dt} \right) + \dfrac{1}{2} \left(\dfrac{di_b^{(u2)}}{dt} - \dfrac{di_b^{(u0)}}{dt} \right) \\[2mm]
+ \dfrac{1}{2} \left(\dfrac{di_a^{(u2)}}{dt} - \dfrac{di_a^{(u0)}}{dt} \right)
\end{array}
\right]
\tag{27}
$$

	p_a	p_b	p_c
$\underline{U}_1$ $+$ $\underline{U}_0$	$1+c_1(\dfrac{di_b^{(u1)}}{dt}-\dfrac{di_b^{(u0)}}{dt})$	$-2+c_1(\dfrac{di_a^{(u1)}}{dt}-\dfrac{di_a^{(u0)}}{dt})$	$1+c_1(\dfrac{di_c^{(u1)}}{dt}-\dfrac{di_c^{(u0)}}{dt})$
$\underline{U}_2$ $+$ $\underline{U}_0$	$-2-c_1(\dfrac{di_c^{(u2)}}{dt}-\dfrac{di_c^{(u0)}}{dt})$	$1-c_1(\dfrac{di_b^{(u2)}}{dt}-\dfrac{di_b^{(u0)}}{dt})$	$1-c_1(\dfrac{di_a^{(u2)}}{dt}-\dfrac{di_a^{(u0)}}{dt})$
$\underline{U}_3$ $+$ $\underline{U}_0$	$1+c_1(\dfrac{di_a^{(u3)}}{dt}-\dfrac{di_a^{(u0)}}{dt})$	$1+c_1(\dfrac{di_c^{(u3)}}{dt}-\dfrac{di_c^{(u0)}}{dt})$	$-2+c_1(\dfrac{di_b^{(u3)}}{dt}-\dfrac{di_b^{(u0)}}{dt})$
$\underline{U}_4$ $+$ $\underline{U}_0$	$1-c_1(\dfrac{di_b^{(u4)}}{dt}-\dfrac{di_b^{(u0)}}{dt})$	$-2-c_1(\dfrac{di_a^{(u4)}}{dt}-\dfrac{di_a^{(u0)}}{dt})$	$1-c_1(\dfrac{di_c^{(u4)}}{dt}-\dfrac{di_c^{(u0)}}{dt})$
$\underline{U}_5$ $+$ $\underline{U}_0$	$-2-c_1(\dfrac{di_c^{(u5)}}{dt}-\dfrac{di_c^{(u0)}}{dt})$	$1+c_1(\dfrac{di_b^{(u5)}}{dt}-\dfrac{di_b^{(u0)}}{dt})$	$1+c_1(\dfrac{di_a^{(u5)}}{dt}-\dfrac{di_a^{(u0)}}{dt})$
$\underline{U}_6$ $+$ $\underline{U}_0$	$1-c_1(\dfrac{di_a^{(u6)}}{dt}-\dfrac{di_a^{(u0)}}{dt})$	$1-c_1(\dfrac{di_c^{(u6)}}{dt}-\dfrac{di_c^{(u0)}}{dt})$	$-2-c_1(\dfrac{di_b^{(u6)}}{dt}-\dfrac{di_b^{(u0)}}{dt})$

Table 1. Definition of position scalars of all voltage vectors in a delta-connected IM.

$$p_\beta = \frac{\sqrt{3}}{2}c_1\left[\left(\frac{di_a^{(u2)}}{dt}-\frac{di_a^{(u0)}}{dt}\right)-\left(\frac{di_b^{(u2)}}{dt}-\frac{di_b^{(u0)}}{dt}\right)\right] \qquad (28)$$

whereby the position can be derived through the arctan operation as shown in (29).

$$\theta_{an} = \arctan\left(p_\beta/p_\alpha\right) \qquad (29)$$

Such a combination between two adjacent voltage vectors, and a null vector, also exists in other five sectors. Therefore, the position estimation is achievable using only the fundamental PWM sequence. **Table 1** gives the position scalars corresponding to all the vectors.

3. Experimental implementation

3.1. Practical considerations

It can be seen that the correct position estimation relies on the precise measurement of the di/dt signals. For this aim, air-cored Rogowski [9], ferrite-cored Rogowski [18], or air-cored coaxial cable-typed [14] transducers can be used. Or direct digital calculation, di/dt = i(t$_2$)−i(t$_1$)/(t$_2$−t$_1$), can be employed instead. **Figure 3** shows a typical di/dt signal along with ADC trigger signal when an air-cored coaxial cable-typed transducer is used.

Another important issue comes from the fact that due to the common/differential mode voltages produced by the inverter, high-frequency oscillation exists in the phase currents, which poses a challenge for accurate di/dt measurement. This is true when dwell times of the voltage vectors are too short or when the reference voltage vectors pass the boundaries of sectors. Therefore, a minimum dwell time, tmin, is imposed on voltage vectors for di/dt measurement. When the original dwell time of a voltage vector, t, is shorter than tmin, an opposite vector with a dwell time, tmin- t, is added to maintain the volt-second. This procedure can be realized simply by the edge-shifting technique, which is illustrated in **Figure 4** when the reference voltage lies in Sector I.

3.2. System schematic

Figure 5 shows the system schematic of the sensorless speed control. Three di/dt sensors are connected in series with the lines of the induction motor, whose parameters are given in **Table 2**. The position vector formation block synthesizes the position vector $p_{\alpha\beta}$ according to the measured di/dts and the sector index (S_i) of the reference voltage vector. For example,

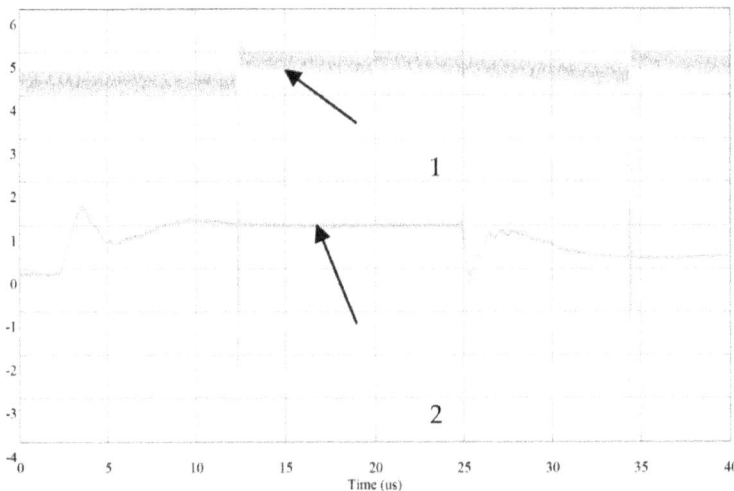

Figure 3. ADC trigger signal (1,V) and di/dt signal (2,V) measured by a air-cored coaxial cable type transducers.

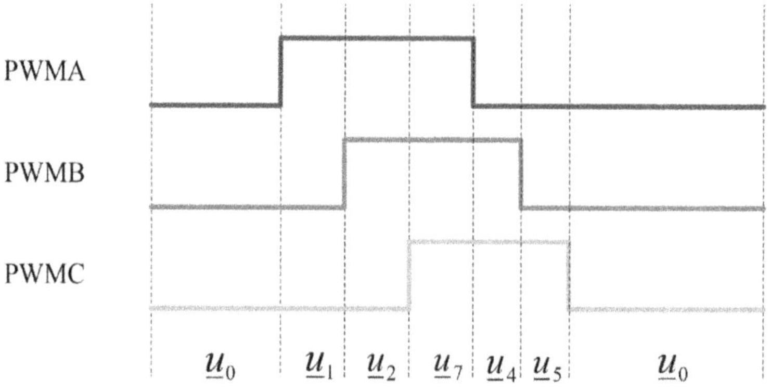

Figure 4. Edge shifting of PWM waveforms when the dwell times of u1 and u2 are smaller than *tmin*.

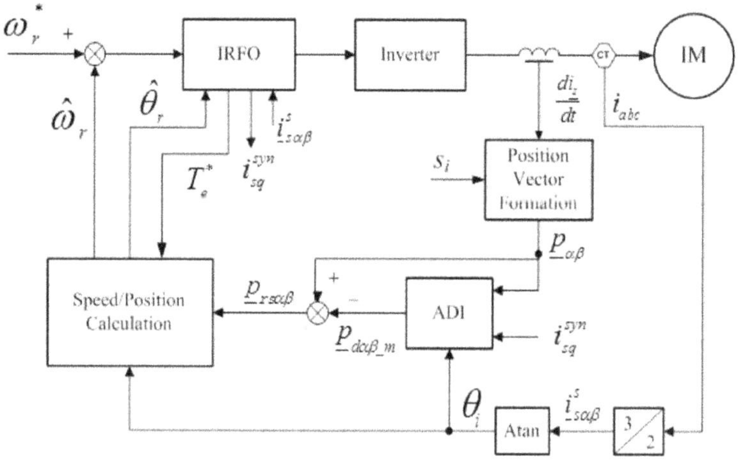

Figure 5. System schematic of the sensorless speed control (ADI).

when $S_i = 1$, Eqs. (27) and (28) can be used. Each of the other five sectors has its own set of equations like (27) and (28) such that the continuous position signal $p_{\alpha\beta}$ can be constructed. Ideally, $p_{\alpha\beta}$ can be used for position/speed estimation, but due to the presence of disturbance signals, which are mainly due to the saturation of the machine, such as the $2f_e$ and $4f_e$ components [3], the estimated position will become distorted if $p_{\alpha\beta}$ is used directly. This disturbance cannot be eliminated by filtering since the disturbance frequencies converge to the wanted signal frequencies as the excitation frequency approaches zero. As a result, a memory-type filter is required. In this work, an adaptive disturbance identifier (ADI) [17] is employed to separate these disturbance signals, i.e., $p_{d\alpha\beta\_m}$, which are filtered out from $p_{\alpha\beta}$. This identifier

Machine Parameters	Values
Rated Power	30kW
Number of Poles	4
Connection	Delta
Lo,Ls	82.5mH, 83.5mH
Rotor Slot Number Nr	56 (unskewed, open slot)
Stator Slot Number	48
Rated Frequency	50 Hz

Table 2. Parameters of the IM.

does not require a pre-commissioning phase. Rather it initiates a learning type sequence in which the resolver disturbance signals are estimated and continually refined as the machine passes through the appropriate torque-speed space [17]. The $p_{d\alpha\beta\_m}$ disturbance signals are stored in memory and are subtracted from the uncompensated signals to give the compensated signal $p_{rs\alpha\beta}$. The resulting position vector $p_{rs\alpha\beta}$ is fed to the Speed/Position Calculation block for the speed and position estimation.

The Speed/Position Calculation block further refines the position signals. Because the ADI-compensated rotor slot position still consists of a speed dependent disturbance signal rotating at $N_r\omega_r/p-2\omega_i$, which is attributed to the inter-modulation effect between the slotting and saturation in the machine [11], and can be conveniently removed by a side band filter [11]. The filtered position signal is shown in **Figure 6**.

The position signals obtained so far can be used for position acquisition in the way shown in (29). However, a mechanical observer similar to [19] is utilized to reduce the noise. The schematic is given in **Figure 7**.

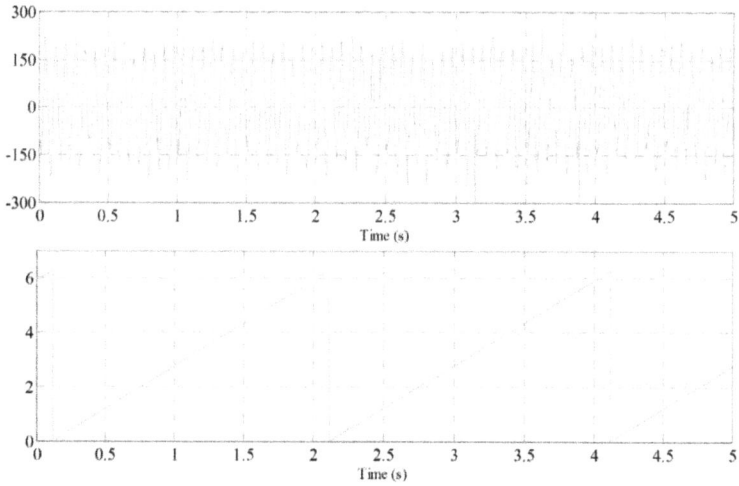

Figure 6. Position estimation at 15 rpm with 75% rated load (*fe* =1.0 Hz). Top: filtered position signal; and bottom: estimated rotor position (rad).

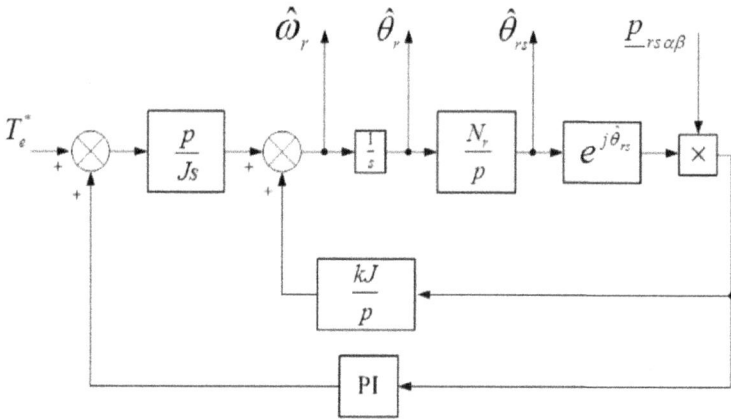

Figure 7. Schematic of the modified mechanical observer (*J* is the moment of inertial and *k* is a design constant, s = d/dt).

Figure 8 shows an improved estimated rotor position signal before and after the observer. An offset angle can be observed between these two rotor positions. This is due to the fact that the mechanical observer's estimation was aligned to the encoder's angle initially, whereas the estimate from the SBF only yields the incremental position.

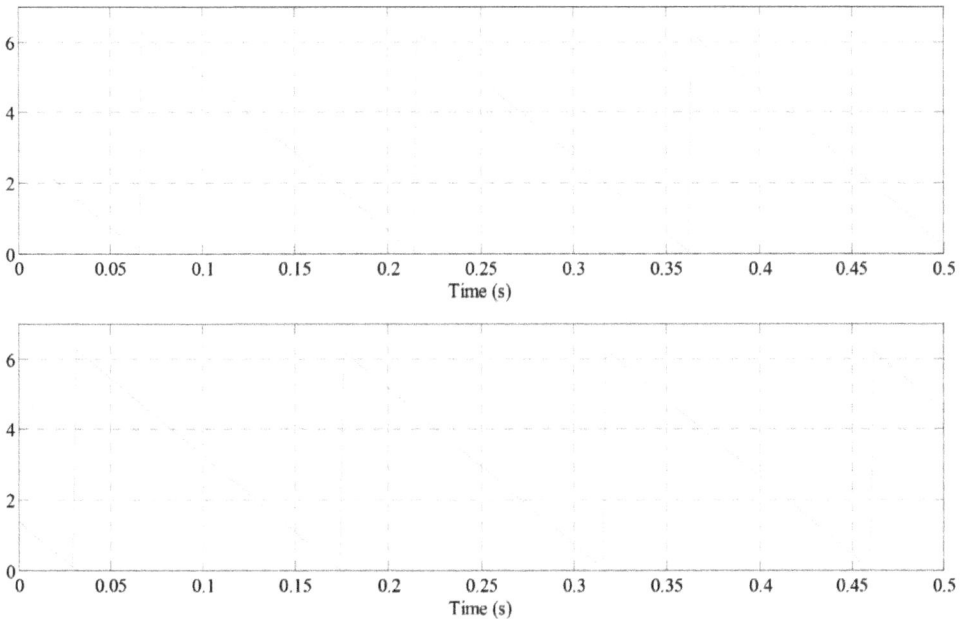

Figure 8. Rotor position estimation (rad) at −210 rpm, no load top: before the observer, bottom: after the observer.

4. Experimental results

The following experiments show the operation of the drive under sensorless speed control at low and higher speeds. The induction motor drive is under speed sensorless control. The load machine is under torque control. The rating of the dynamometer converter is such that only a loading of 80% rate can be applied to the induction motor. It is emphasized that all the experimental waveforms were recorded on an experiment rig.

Figure 9 shows that under no load condition the IM is reversed at 6 rpm, corresponding to the excitation frequency f_e = 0.2 Hz. Good rotor estimation can be seen. In **Figure 10**, the drive performs a speed reversal at ±12 rpm under 70% rated load condition. At −12 rpm, this load condition corresponds to the drive under braking at zero excitation frequency. Since a constant torque is applied to the DC load machine, under speed reversal, the amplitude of i_{sq} changes due to the losses involved.

In **Figure 11**, the drive was taken between ±210 rpm under no load. This test confirms the capability of the drive to perform larger speed transients.

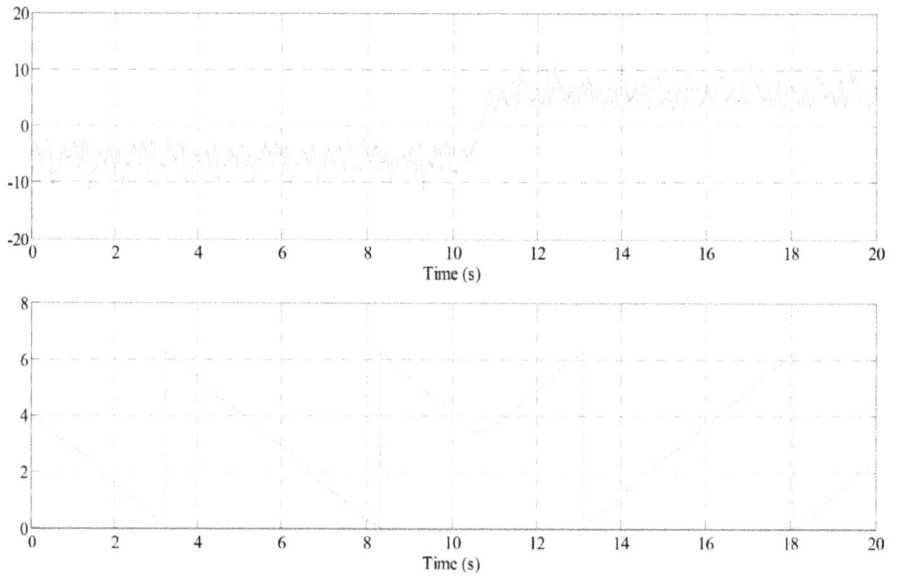

Figure 9. Speed reversal at 6 rpm, no load top: measured speed (rpm), bottom: estimated rotor position (rad).

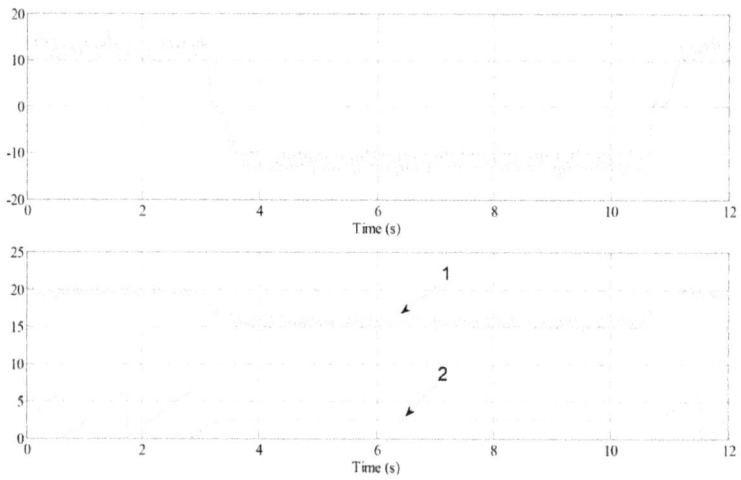

Figure 10. Speed reversal at ±12 rpm under 70% rated load top: measured speed (rpm), bottom: rotor flux angle (1, rad) and filtered Isq (2, A).

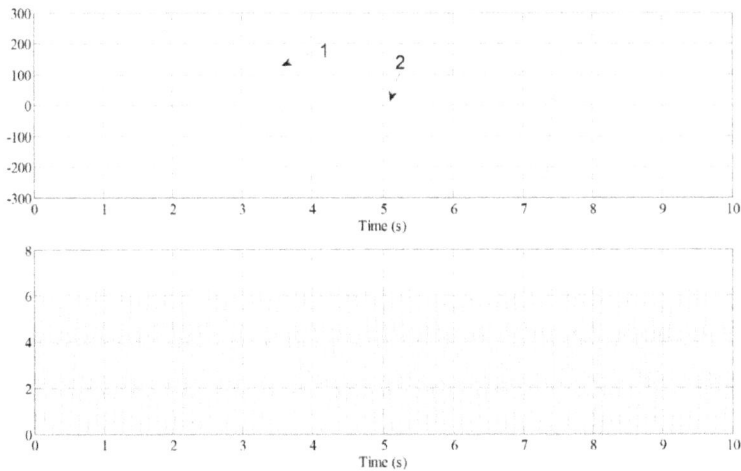

Figure 11. Speed reversal at ±240 rpm, no load top: measured speed (1, rpm) and filtered Isq (2, A); bottom: estimated rotor position (rad).

5. Conclusion

A sensorless position scheme for AC machines is presented relying on the line-current derivative measurements in response to a fundamental PWM switching sequence. The rotor position angle is derived due to the tracking of rotor slotting and the signal is used for sensorless control. If the anisotropy caused by the main flux saturation is tracked, for example, in a permanent magnet machine, this method is applicable throughout a very wide speed range. If rotor slotting is tracked, then the maximum speed is limited by the Nyquist frequency associated with the rotor slot passing frequency. In principle, however, the method can work over the entire torque-speed envelope.

Author details

Qiang Gao[1]*, Greg Asher[2] and Mark Sumner[2]

*Address all correspondence to: gaoqiang@sjtu.edu.cn

1 Shanghai Jiao Tong University, Shanghai, China

2 University of Nottingham, Nottingham, UK

References

[1] Jansen PL, Lorenz RD. Transducerless field orientation concepts employing saturation induced saliencies in induction machines. In: Proceedings of the IEEE IAS Annual Meeting; Nov/Dec 1995. p. 174-181

[2] Holtz J. Sensorless position control of induction motors – An emerging technology. IEEE Transactions on Industrial Electronics. 1998;45(6):840-852. DOI: 10.1109/41.735327

[3] Teske N, Asher GM, Sumner M, Bradley KJ. Encoderless position estimation for symmetric cage induction machines under loaded conditions. IEEE Transactions on Industrial Applications. 2001;37(6):1793-1800. DOI: 10.1109/28.968193

[4] Silva CA, Asher GM, Sumner M, Bradley KJ. Sensorless rotor position control in a surface mounted PM machine using HF voltage injection. In: Proceedings of the EPE –PEMC on CD-ROM; 2002

[5] Linke M, Kennel R, Holtz J Sensorless position control of permanent magnet synchronous machines without limitation at zero speed. In: Proceedings of the IEEE IECON 2002 on CD-ROM; 2002

[6] Corley MJ, Lorenz RD. Rotor position and velocity estimation for a salient-pole permanent magnet synchronous machine at standstill and high speeds. IEEE Transactions on Industry Applications. 1998;34(4):784-789. DOI: 10.1109/28.703973

[7] Ha JI, Sul SK. Sensorless field orientation of an induction machine by high frequency signal injection. In: Proceedings of the IEEE IAS Annual Meeting; 1997. p. 426-432

[8] Schroedl M. Sensorless control of AC machines at low speed and standstill based on the INFORM method. In: Proceedings of the IEEE IAS Annual Meeting; 1996. p. 270-277

[9] Caruana C, Asher GM, Clare J. Sensorless flux position estimation at low and zero frequency by measuring zero-sequence current in delta connected cage induction machines. In: Proceedings of the IEEE IAS Annual Meeting on CD-ROM; 2003

[10] Staines CS, Asher GM, Sumner M. Sensorless control of induction machines at zero and low frequency using zero sequence currents. In: Proceedings of the IEEE IAS Annual Meeting on CD-ROM; 2004

[11] Holtz J, Pan H. Elimination of saturation effects in sensorless position controlled induction motors. In: Proceedings of the IEEE IAS Annual Meeting on CD-ROM; 13–18 October 2002; Pittsburgh

[12] Robeischl E, Schroedl M. Optimized INFORM measurement sequence for sensorless PM synchronous motor drives with respect to minimum current distortion. IEEE Transactions on Industrial Applications. 2004;40(2):591-598. DOI: 10.1109/TIA.2004.824510

[13] Wolbank T, Machl J. A modified PWM scheme in order to obtain spatial information of ac machines without mechanical sensor. In: Proceedings of the IEEE APEC; 2002. p. 310-315

[14] Juliet J, Holtz J. Sensorless Acquisition of the rotor position angle for induction motors with arbitrary stator windings. In: Proceedings of the IEEE IAS Annual Meeting on CD-ROM; 2004

[15] Erdman JM, Kerkman RJ, Schlegel DW, Skibinski GL. Effect of PWM inverters on AC motor bearing currents and shaft voltages. IEEE Transactions on Industrial Applications. 1996;**32**(2):250-259. DOI: 10.1109/28.491472

[16] Gao Q, Asher GM, Sumner M, Makyš P. Position estimation of AC machines at all frequencies using only space vector PWM based excitation. In: Proceedings of the IEE 3rd International Conference on Power Electronics, Machines and Drives (PEMD); 2006. p. 61-70

[17] Gao Q, Asher GM, Sumner M. Sensorless position and speed control of induction motors using high frequency injection and without off-line pre-commissioning. In: Proceedings of the 31st Annual Meeting of IEEE, IES 2005 on CD-ROM; 2005

[18] Wolbank TM, Machl JL, Hauser H. Closed-loop compensating sensors versus new current derivative sensors for shaft-sensorless control of inverter fed induction machines. IEEE Transactions on Instrumentation and Measurement. 2004;**53**(4):1311-1315. DOI: 10.1109/TIM.2004.830561

[19] Jansen PL, Lorenz RD. Transducerless position and velocity estimation in induction and salient AC machines. IEEE Transactions on Industry Applications. 1995;**31**(2):240-247. DOI: 10.1109/28.370269

Rotor Flux Reference Generation Control Strategy for Direct Torque Controlled DFIG

Gopala Venu Madhav and Y. P. Obulesu

Additional information is available at the end of the chapter

http://dx.doi.org/10.5772/intechopen.75362

Abstract

The wind turbines based Doubly Fed Induction Generator (DFIG) is not able to support the voltage and the frequency of the grid during and immediately following the grid failure. This would cause major problems for the systems stability, but the turbines should stay connected to the grid in case of a failure. This can be achieved by using crowbar protection in particularly during voltage dips. When low depth voltage dips occur, the necessity of crowbar protection can be eliminated by using proposed Direct Torque Control (DTC), with a proper rotor flux generation strategy, by which during the fault it will be possible to maintain the machine connected to grid, generating power from the wind, reducing the stator and rotor over currents, eliminating the torque oscillations that normally produce such voltage dips and fast dynamic response accompanies the overall control of the wind turbine. In this chapter, the DFIG performance is analyzed and the results are presented for with proposed control strategy with and without voltage dip, without control strategy with voltage dip, and control strategy during longer voltage dip.

Keywords: crowbar protection, direct torque control, doubly fed induction generator, reference generation strategy, voltage dip

1. Introduction

The main objective of the control strategy proposed for DFIG [1] in this chapter is to eliminate the necessity of the crowbar protection [2] when low voltage dips occur. Hence, by using Direct Torque Control (DTC), with a proper rotor flux generation control strategy, during the fault it is possible to maintain the machine connected to the grid [3, 4], generating power from the wind, reducing over currents, and eliminating the torque oscillations that normally produce over voltage dips [5–9]. Elimination of crowbar protection is cost effective and reducing bulkiness and reducing circuitry besides the above mentioned advantages of the proposed strategy.

This proposed control strategy gives fast dynamic response for voltage dips during electrical grid disturbances keeping the stator and rotor over currents within the considerable limits.

In Section 1, essence of the rotor flux reference generation scheme with Direct Torque Control (DTC) strategy of DFIM is briefly given.

In Section 2, description of the proposed control strategy is given.

In Section 3, the rotor flux reference generation strategy is described.

In Section 4, the results are presented for with and without fault condition for low and longer voltage dip.

In Section 5, the summary of the chapter is given.

2. Direct torque control strategy for rotor side converter

Figure 1 shows the wind turbine generation system with the DTC technique along with the rotor flux amplitude reference generation strategy to control the DFIG during the unbalanced

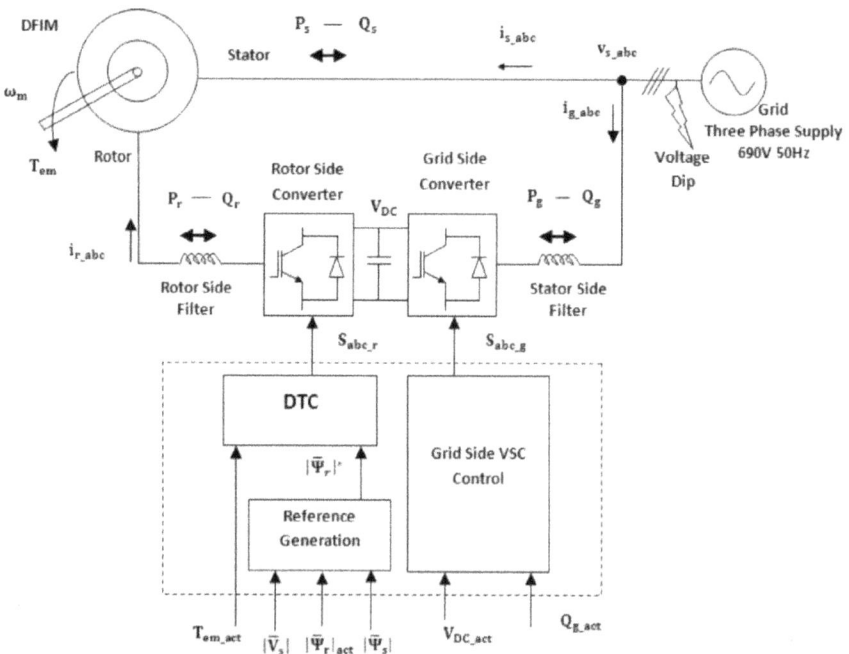

Figure 1. Block diagram of DTC of DFIM with rotor flux reference generation control strategy.

condition, i.e., during an voltage dip. During the voltage dip, if DFIG is maintained with constant electromagnetic torque and rotor flux amplitude, that means if no control strategy is been adopted then it leads to non-sinusoidal grid currents making the grid to be in unstabilized condition. The proposed control strategy eliminates the perturbations in electromagnetic torque, makes it to be within the stabilized limits, reduce the stator and rotor overcurrents produced leading to elimination of the crowbar protection during low voltage dips and generate sinusoidal grid currents without the necessity to change the hardware requirement and also the prevalent control philosophy adopted. The behavior of the DFIG during the voltage dip with and without proposed control strategy is validated with the results presented.

The DFIM is fed with back-to-back converter. It consists of two converters, i.e., machine-side converter and grid-side converter that are connected "back-to-back." Between the two converters a dc-link capacitor is placed, as energy storage, in order to keep the voltage variations (or ripple) in the dc-link voltage small. With the machine-side converter it is possible to control the torque or the speed of the DFIG and also the power factor at the stator terminals, the main objective for the grid-side converter is to keep the dc-link voltage constant. As the rotor current or voltage is lower, power is lower because of which the converter rating is 30% of the full-rated machine which makes to be the main advantage of DFIM.

In this chapter, a control strategy has been developed for the rotor side converter to generate rotor flux reference. The conventional Pulse Width Modulation (PWM) technique is adopted for grid-side converter; the converter maintains the dc-link voltage to be constant and also supplies the reactive power to the grid through it. As shown in **Figure 1**, the DFIM control is divided into two different control blocks. A DTC that controls the machine's torque (T_{em}) and the rotor flux amplitude ($|\overline{\psi}_r|$) with high dynamic capacity, and a second block that generates the rotor flux amplitude reference, in order to handle with the voltage dips. The details of rotor flux reference generation are shown in **Figure 2**. The required rotor voltage vector is selected based on the vector selection table as mentioned in **Table 1**.

When the wind turbine is affected by a voltage dip, it needs to address three main problems: the first problem is based on the view of the control strategy being adopted, the dip produces control difficulties, since it is a perturbation in the winding of the machine that is not being directly

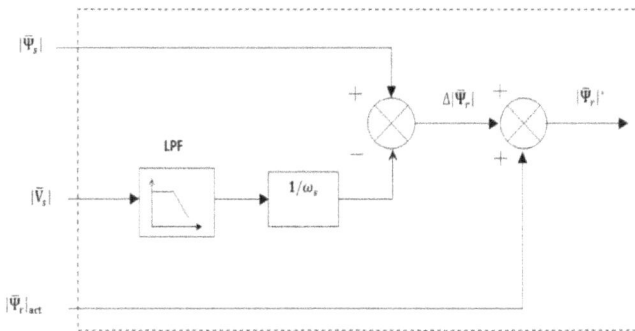

Figure 2. Details of rotor flux reference generation control strategy.

	Error of electromagnetic torque		
	1	0	−1
Error of rotor flux 1	$V_{(n-1)}$	V_0, V_7	$V_{(n+1)}$
_1	$V_{(n-2)}$	V_0, V_7	$V_{(n+2)}$

n = sector

Table 1. Selection of voltage vectors.

controlled (the stator); the second problem is the dip generates a disturbance in the stator flux, making necessary higher rotor voltage to maintain control on the machine currents; and the third problem is if there are no special improvements being adopted, the power delivered through the rotor by the back-to-back converter, will be increased due to the increase of voltage and currents [3, 5] in the rotor of the machine, provoking finally, an increase of the dc bus voltage [10, 11].

Taking into account this, depending on the dip depth and asymmetry, together with the machine operation conditions at the moment of the dip (speed, torque, mechanical power, etc.), implies that the necessity of the crowbar protection is inevitable in many faulty situations [12]. However, in this chapter, a control strategy that eliminates the necessity of the crowbar activation in some low depth voltage dips is proposed.

3. Rotor flux reference generation control strategy

The stator flux evolution of the machine is determined from the stator voltage equation as given by:

$$\overline{v}_s^s = R_s \overline{i}_s^s + \frac{d\overline{\psi}_s^s}{dt} \tag{1}$$

The torque can be estimated by using the following expression:

$$T_{em} = \frac{3}{2} P \frac{L_m}{\sigma L_s L_r} \text{Im}\left\{ \overline{\psi}_r^* \cdot \overline{\psi}_s^s \right\} \tag{2}$$

From the Eq. (1), considering the stator resistive drop as negligible, the unbalance in the grid voltage will directly affect the stator voltage and because of that it affects the stator flux space vector as the stator is directly connected to grid. That means, the oscillating behavior produced in the grid voltage due to unbalance will be reflected onto the stator flux space vector and further onto the rotor flux space vector [5]. The unbalance case of both the stator and the rotor flux space vectors can be represented mathematically as:

$$\overline{\psi}_s = \psi_{\alpha s} + j\psi_{\beta s} = \overline{\Psi}_{\alpha s}\cos(\omega t + \delta) + j\overline{\Psi}_{\beta s}\sin(\omega t + \delta)$$
$$\overline{\psi}_r = \psi_{\alpha r} + j\psi_{\beta r} = \overline{\Psi}_{\alpha r}\cos(\omega t) + j\overline{\Psi}_{\beta r}\sin(\omega t) \tag{3}$$

Substituting Eq. (3) in Eq. (2), it leads to:

$$T_{em} = \frac{3}{4}p\frac{L_m}{\sigma L_s L_r}\left[\left(\overline{\Psi}_{\beta s}\overline{\Psi}_{\alpha r} + \overline{\Psi}_{\alpha s}\overline{\Psi}_{\beta r}\right)\sin(\delta) + \left(\overline{\Psi}_{\beta s}\overline{\Psi}_{\alpha r} - \overline{\Psi}_{\alpha s}\overline{\Psi}_{\beta r}\right)\sin(2\omega t + \delta)\right] \quad (4)$$

From Eq. (4), it can be observed that the torque expression consists of a constant term and an oscillating term. In general, for a given machine torque should be constant and it should not be oscillatory, if it is so it will lead to mechanical instability of the wind energy conversion system. Therefore, the oscillatory term has to be somehow canceled or make it zero. So, equating the oscillatory term to zero, it leads to condition, i.e., Eq. (5), wherein the ratio of the amplitudes of rotor and stator flux space vectors should be equal, which has to be maintained properly during the unbalanced condition. Otherwise, it will lead to oscillatory behavior of stator and rotor fluxes and even the currents.

$$\frac{\overline{\Psi}_{\alpha r}}{\overline{\Psi}_{\beta r}} = \frac{\overline{\Psi}_{\alpha s}}{\overline{\Psi}_{\beta s}} \quad (5)$$

From Eq. (5), one more inference is that the rotor flux reference generation should be in accordance to the above equation. Further, it can be deduced that during the unbalance condition as the stator flux space vector oscillates, likewise the rotor flux space vector should be made oscillatory, so that the torque with respect to the Eq. (2) is constant and sinusoidal currents exchange with the grid. If otherwise, it leads to oscillatory behavior in torque and leads to non-sinusoidal currents exchange with the grid.

As said previously, from Eq. (1), the unbalance grid voltage will produce an oscillatory behavior in stator flux space vector and further in rotor flux space vector. This oscillatory behavior in terms of the amplitudes of both stator and rotor flux space vectors similar to the unbalance voltage space vector [5] can be expressed as:

$$|\overline{\Psi}_s|^2 = \left[\frac{\overline{\Psi}_{\alpha s}^2 + \overline{\Psi}_{\beta s}^2}{2}\right] + \left[\frac{\overline{\Psi}_{\alpha s}^2 - \overline{\Psi}_{\beta s}^2}{2}\right]\cos(2\omega t + \delta)$$

$$|\overline{\Psi}_r|^2 = \left[\frac{\overline{\Psi}_{\alpha r}^2 + \overline{\Psi}_{\beta r}^2}{2}\right] + \left[\frac{\overline{\Psi}_{\alpha r}^2 - \overline{\Psi}_{\beta r}^2}{2}\right]\cos(2\omega t)$$

$$(6)$$

Eq. (6) fulfill the Eq. (5), so in order to produce constant torque and sinusoidal currents to be exchanged with the grid, the rotor flux reference generation should be according to Eq. (6). Further discussion shows how this oscillatory rotor flux reference generation is created.

Stator flux equations are given in Eq. (7) (neglecting stator resistance, Rs) [1], it is approximated as the addition of exponential and sinusoidal term.

$$\overline{\Psi}_{\alpha s} = K_1 e^{-K_2 t} + K_3\cos(\omega_s t + K_4)$$
$$\overline{\Psi}_{\beta s} = K_5 e^{-K_2 t} + K_3\sin(\omega_s t + K_4)$$

$$(7)$$

where K_1 to K_5 are constants which depends on nature and the moment when voltage dip occurs. In accordance to Eq. (5), the exponential term in Eq. (7) can be eliminated by producing simultaneous oscillations in rotor flux as produced in stator flux due to unbalance.

The stator and rotor currents are given in Eq. (8).

$$\bar{i}_s^s = \frac{L_m}{\sigma L_r L_s}\left(\frac{L_r}{L_h}\overline{\Psi}_s^s - \overline{\Psi}_r^s\right)$$

$$\bar{i}_r^s = \frac{L_m}{\sigma L_r L_s}\left(\frac{L_s}{L_h}\overline{\Psi}_r^s - \overline{\Psi}_s^s\right) \tag{8}$$

As depicted in **Figure 2**, the proposed rotor flux amplitude reference generation strategy, adds a term $(\Delta|\overline{\Psi}_r|)$ to the required reference rotor flux amplitude according to the following expression:

$$\Delta|\overline{\Psi}_r| = |\overline{\Psi}_s| - \frac{|\overline{v}_s|}{\omega_s} \tag{9}$$

with $|\overline{\Psi}_s|$, the estimated stator flux amplitude and $|\overline{v}_s|$ voltage of the grid (not affected by the dip). This voltage can be calculated by several methods, for instance, using a simple small bandwidth low-pass filter, as illustrated in **Figure 2**. It must be highlighted that constants K_1–K_5 from Eq. (7) are not needed in the rotor flux reference generation reducing its complexity.

The stator and rotor fluxes and their magnitudes can be calculated by using:

$$\overline{\Psi}_s = L_s\bar{i}_s + L_m\bar{i}_r$$

$$\overline{\Psi}_r = L_m\bar{i}_s + L_r\bar{i}_r$$

$$|\Psi_s| = \sqrt{\Psi_{ds}^2 + \Psi_{qs}^2} \tag{10}$$

$$|\Psi_r| = \sqrt{\Psi_{dr}^2 + \Psi_{qr}^2}$$

When there is voltage dip condition, the proposed control scheme makes the rotor flux produce the oscillations in similar to stator flux and when there is no dip the stator and rotor fluxes will be constant, which means the term $\Delta|\overline{\Psi}_r|$ shown in **Figure 2** is zero.

4. Results and discussion

The ratings of the DFIM and the wind turbine are 2.6 MW, 690 V, 50 Hz, 4-pole machine and 3 blades, rotor diameter of 70 m, hub height of 84.3 m, cut-in wind speed of 3 ms^{-1}, cut-out wind speed of 25 ms^{-1} and rated wind speed of 15 ms^{-1}, respectively. The stator-to-rotor turns ratio, N_s/N_r is 0.34, and the rotor current is approximately 0.34 times smaller than the stator current, if the magnetizing current is neglected. The stator-to-rotor turns ratio of the DFIG is required to estimate the ohmic loss as it depends on current passing through it.

4.1. Analysis of DFIG with rotor flux reference generation without voltage dip

The results are presented in the case of without symmetrical voltage dip as shown in **Figure 3** that means, the value of $\Delta|\overline{\Psi}_r|$ in **Figure 2** will be zero, therefore the required value of rotor flux will be the reference value of flux. The stator voltage waveform shown in **Figure 3(a)**, from the figure, it is observed that the stator voltage is constant under normal operation.

The torque is maintained at its generated value of 0.2 pu as there is no consideration of voltage dip, which is clearly shown in **Figure 3(b)**.

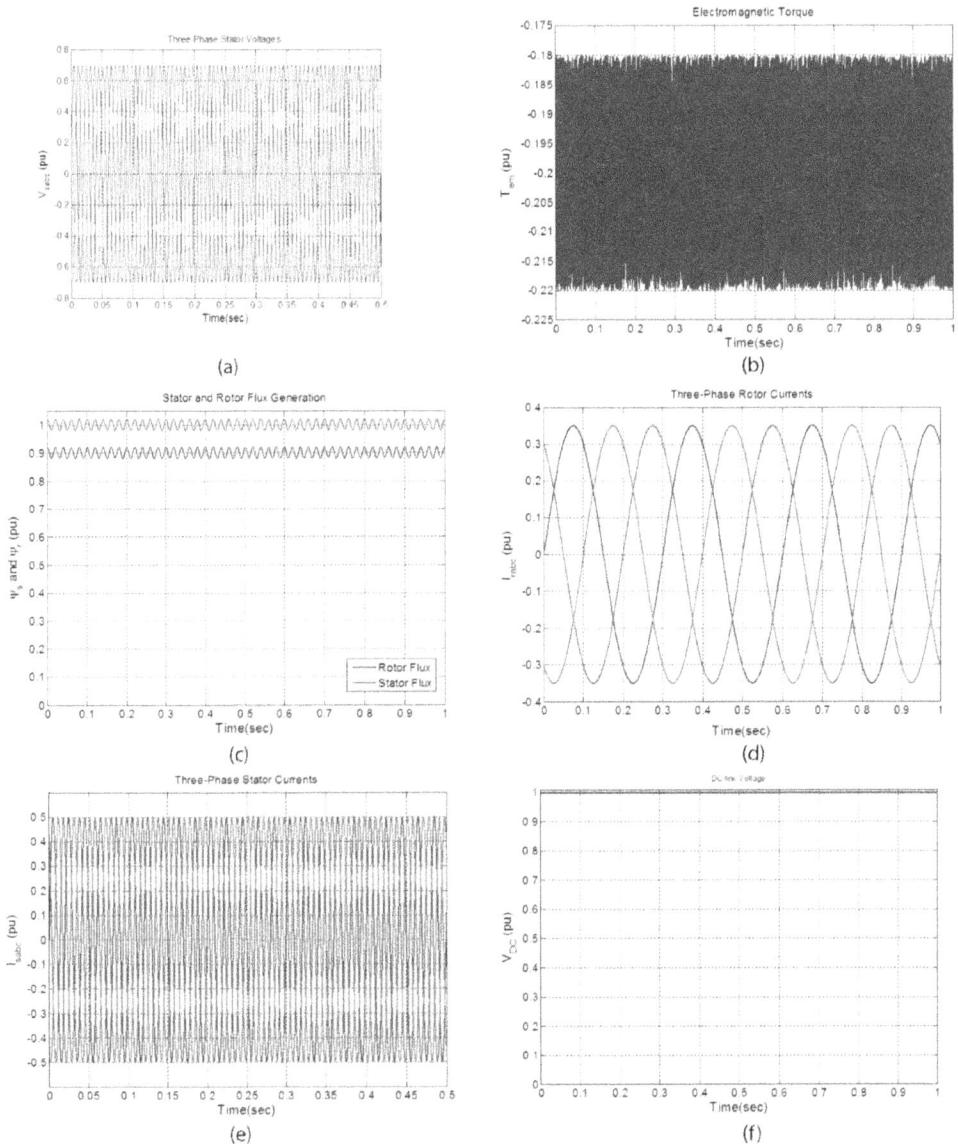

Figure 3. (a) Stator voltages of DFIM with proposed rotor flux reference generation without voltage dip, (b) torque of DFIM with proposed rotor flux reference generation without voltage dip, (c) stator and rotor flux of DFIM with proposed rotor flux reference generation without voltage dip, (d) rotor currents of DFIM with proposed rotor flux reference generation without voltage dip, (e) stator currents of DFIM with proposed rotor flux reference generation without voltage dip, and (f) DC-link voltage of DFIM with proposed rotor flux reference generation without voltage dip.

The responses of the stator and rotor flux without the voltage dip are shown in **Figure 3(c)**. Note that at the steady state, without the presence of a dip, the term $\Delta|\overline{\Psi}_r|$ will be zero in Eq. (9).

Figure 3(d) shows the rotor currents response for without voltage dip, which clearly indicates the steady state operation of DFIM.

Figure 3(e) shows the normal operated result of stator currents under normal operation, that is, without any over currents.

The response of DC-link voltage is shown in **Figure 3(f)**, it is noticed from the figure that the DC-link voltage is maintained constant. When the wind energy system is operating under normal or abnormal condition, the DC-link voltage has to be maintained constant, but to mitigate the over currents in rotor and stator produced due to voltage dip by adopting the proposed control strategy, maintaining the constant value of DC-link voltage is lost. This is explained with the case of voltage dip with and without rotor flux reference generation scheme.

4.2. Analysis of DFIG without rotor flux reference generation with voltage dip

Results are presented for the proposed control strategy to show its effectiveness under low voltage dips, in this case 30%, as illustrated in **Figure 4(a)**, symmetric voltage dip considered with and without the proposed flux reference generation strategy and at nearly constant speed. The symmetrical three phases to ground fault is created at 0.8 s and the fault is cleared off once the time reaches 0.9 s, after which the voltage starts to recover to normal value as shown in **Figure 4(a)**.

From the **Figure 4(b)**, it is observed that the generated torque before the voltage dip is maintained at −0.2 pu. When the dip occurs there are high transient peaks in it, this is because the value of the required rotor voltage is more than the DC-link voltage at that particular instant; otherwise the DTC technique tries to maintain the torque constant when the fault is cleared.

The response of the stator and rotor fluxes is shown in **Figure 4(c)**. As it is the case of without rotor flux reference generation, the rotor flux doesn't follow the stator flux, which can be clearly seen because of which torque has perturbations as can be seen in **Figure 4(b)**.

The response of rotor currents is shown in **Figure 4(d)**. The **Figure 4(d)** clearly shows the high values of rotor currents are produced at the instant of voltage dip, but the DTC technique manages to control the rotor currents still within its limits.

It is observed from the **Figure 4(e)**, high values of stator currents are produced due to abnormal condition, the values of stator currents crosses more than 0.9pu and settles back to steady state value once the fault is cleared.

The DC-link voltage oscillations for without reference generation for short duration of voltage dip can be clearly seen to be balanced and sinusoidal as shown in **Figure 4(f)**. As in this case, there are no special improvements being adopted, the power delivered through the rotor by the ac-dc-ac converter will be increased due to the increase of voltage and currents in the rotor of the DFIM, provoking finally an increase in the DC-link voltage.

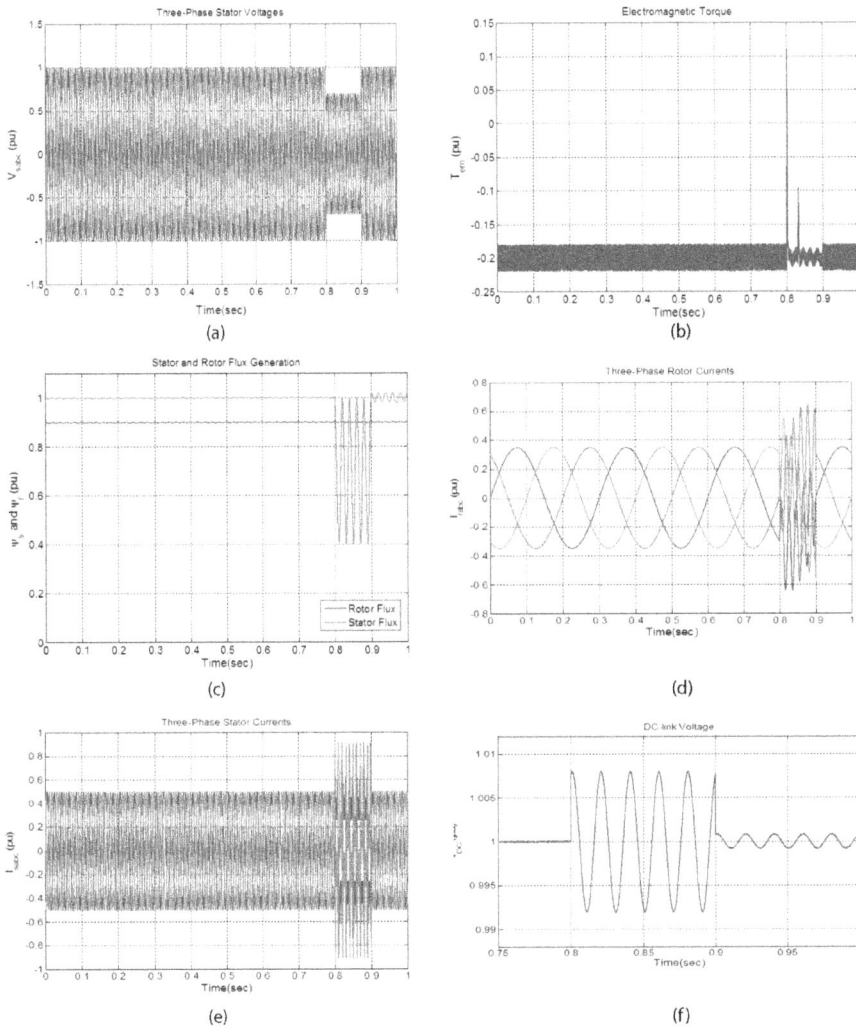

Figure 4. (a) Stator voltages of DFIM without proposed rotor flux reference generation, (b) torque of DFIM without proposed rotor flux reference generation, (c) stator and rotor flux of DFIM without proposed rotor flux reference generation, (d) rotor currents of DFIM without proposed rotor flux reference generation, (e) stator currents of DFIM without proposed rotor flux reference generation, and (f) DC-link voltage of DFIM without proposed rotor flux reference generation.

4.3. Analysis of DFIG with rotor flux reference generation with voltage dip

The response of the generated torque is shown in **Figure 5(a)**. From the **Figure 5(a)**, the high peaks produced in torque are eliminated, which were produced due to dip when without rotor flux reference generation scheme is considered. The high peaks are eliminated by producing the oscillations in rotor flux along with the oscillations produced in stator flux during dip due to poor damped poles. This torque response indicates that the mechanical stresses are reduced on the wind energy system.

The oscillations produced in rotor flux are clearly seen from the **Figure 5(b)**, which follows close to stator flux oscillations. This is achieved by the proposed rotor flux reference generation scheme employed as shown in **Figure 2**.

Consequently, the proposed control scheme maintains the stator and rotor currents under their safety limits, avoiding high over currents, either in the voltage fall or rise. The proposed strategy is analyzed for three phase fault. However, as predicted in theory, it is hard to avoid a deterioration of the quality of these currents. The response of rotor currents of DFIM with proposed scheme is shown in **Figure 5(c)**.

The response of the stator currents of DFIM is shown in **Figure 5(d)**, wherein the stator currents are within the limits when compared to stator currents produced by without rotor flux reference generation scheme as shown in **Figure 4(e)**.

Moreover, by mitigating the over currents of the rotor, the back-to-back converter is less affected by this perturbation, producing short dc bus voltage oscillations. The DC-link voltage oscillations for with rotor flux reference generation are shown in **Figure 5(e)**. The DC-link voltage oscillations are unbalanced but sinusoidal and are constant as shown in **Figure 5(e)**.

4.4. Analysis of DFIG without rotor flux reference generation during longer voltage dip

The results for continuous dip are shown in **Figures 6** and **7** for both without and with reference rotor flux generation respectively. The duration of the longer voltage dip is from 0.2 to 1 s, which can be seen with three phase stator voltage in **Figure 6(a)**.

As showed in **Figure 6(b)**, there are number of perturbations in torque due to exceeding of requirement of rotor voltage compared to the actual DC-link voltage. This causes mechanical stresses on the wind energy conversion system, which is not good for the wind turbine.

The responses of the stator and rotor flux are shown in **Figure 6(c)**, and it is observed from the figure that there are some oscillations in stator flux and no oscillations in rotor flux.

Figure 6(d) shows the response of rotor currents due to longer voltage dip. The rotor currents reach its limits and from the **Figure 6(d)**, it can be clearly seen that there is complete unbalance in the rotor currents but as said they just reach the limits.

The over currents in the stator can be clearly seen in **Figure 6(e)**, due to increase in the rotor currents.

The response of the DC-link voltage with balanced sinusoidal oscillations due to the fault is shown in **Figure 6(f)**.

4.5. Analysis of DFIG with rotor flux reference generation during longer voltage dip

Figure 7(a) clearly shows the torque is maintained at its required value, without the high peaks caused due to longer voltage dip, which allows eliminating mechanical stresses on the wind turbine.

The necessary rotor flux reference generation is to overcome the problems due to the longer voltage dip along with the stator flux oscillations as shown in **Figure 7(b)**.

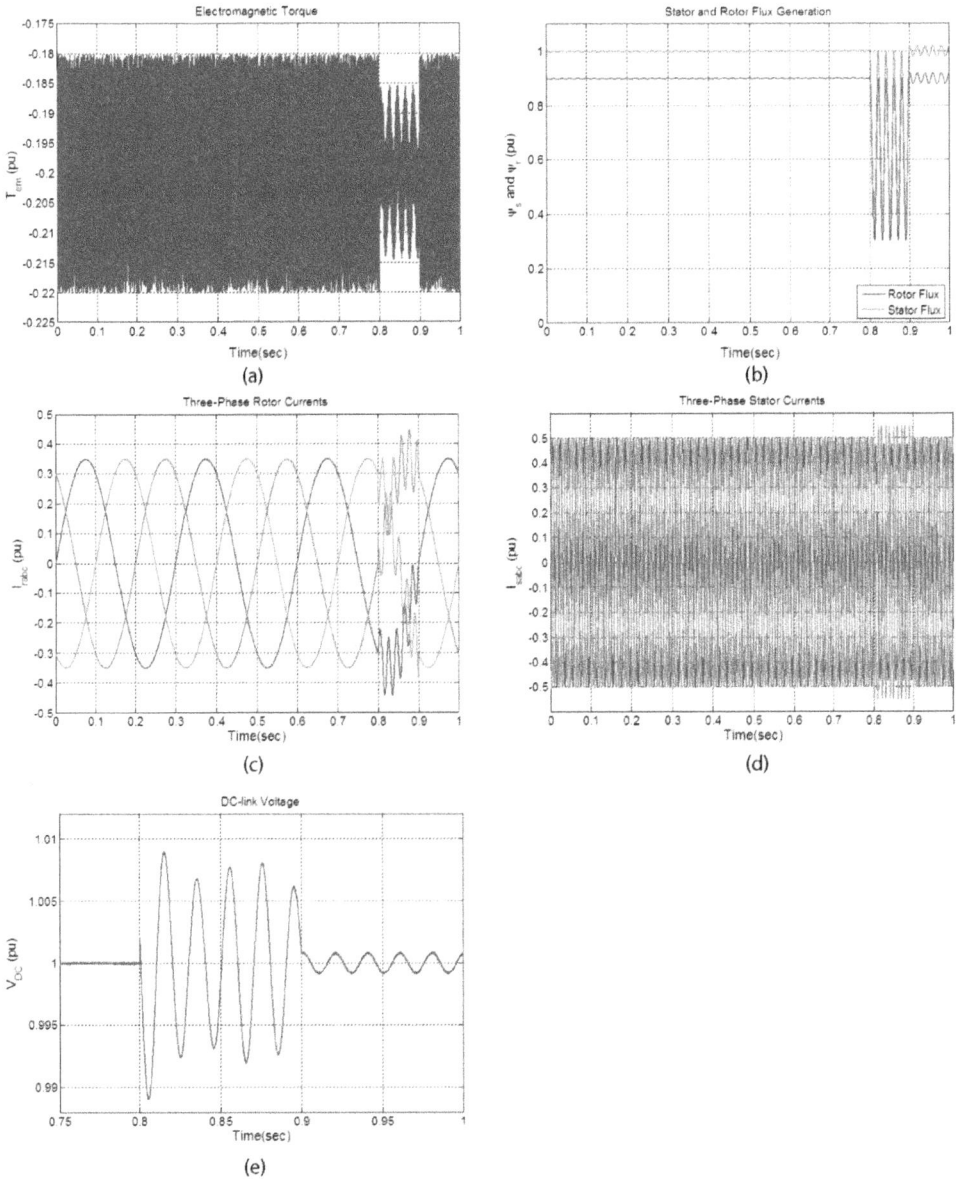

Figure 5. (a) Torque of DFIM with proposed rotor flux reference generation, (b) stator and rotor flux of DFIM with proposed rotor flux reference generation, (c) rotor currents of DFIM with proposed rotor flux reference generation, (d) stator currents of DFIM with proposed rotor flux reference generation, and (e) DC-link voltage of DFIM with proposed rotor flux reference generation.

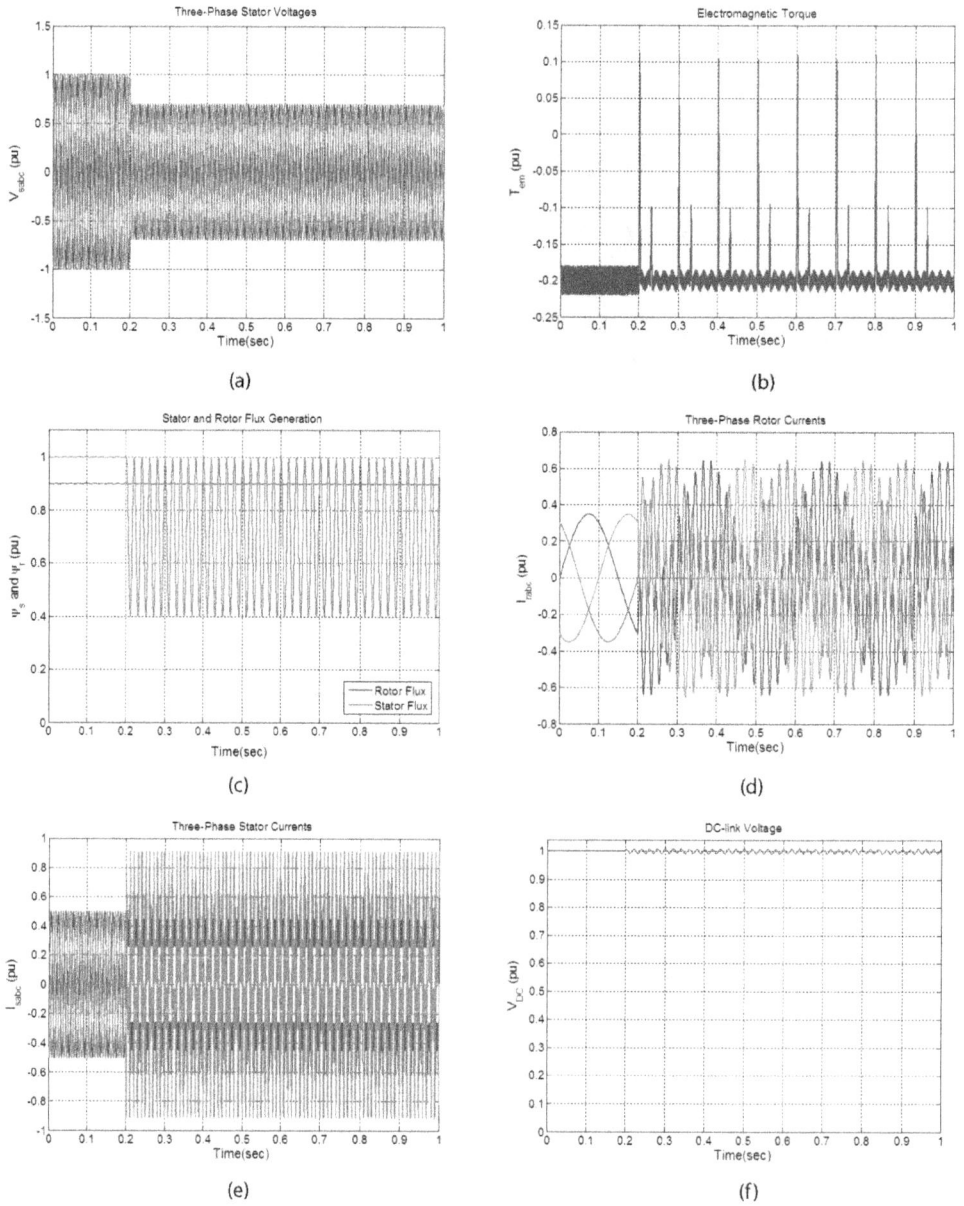

Figure 6. (a) Stator voltages of DFIM without proposed rotor flux reference generation, (b) torque of DFIM without proposed rotor flux reference generation, (c) stator and rotor flux of DFIM without proposed rotor flux reference generation, (d) rotor currents of DFIM without proposed rotor flux reference generation, (e) stator currents of DFIM without proposed rotor flux reference generation, and (f) DC-link voltage of DFIM without proposed rotor flux reference generation.

(a)

(b)

(c)

(d)

(e)

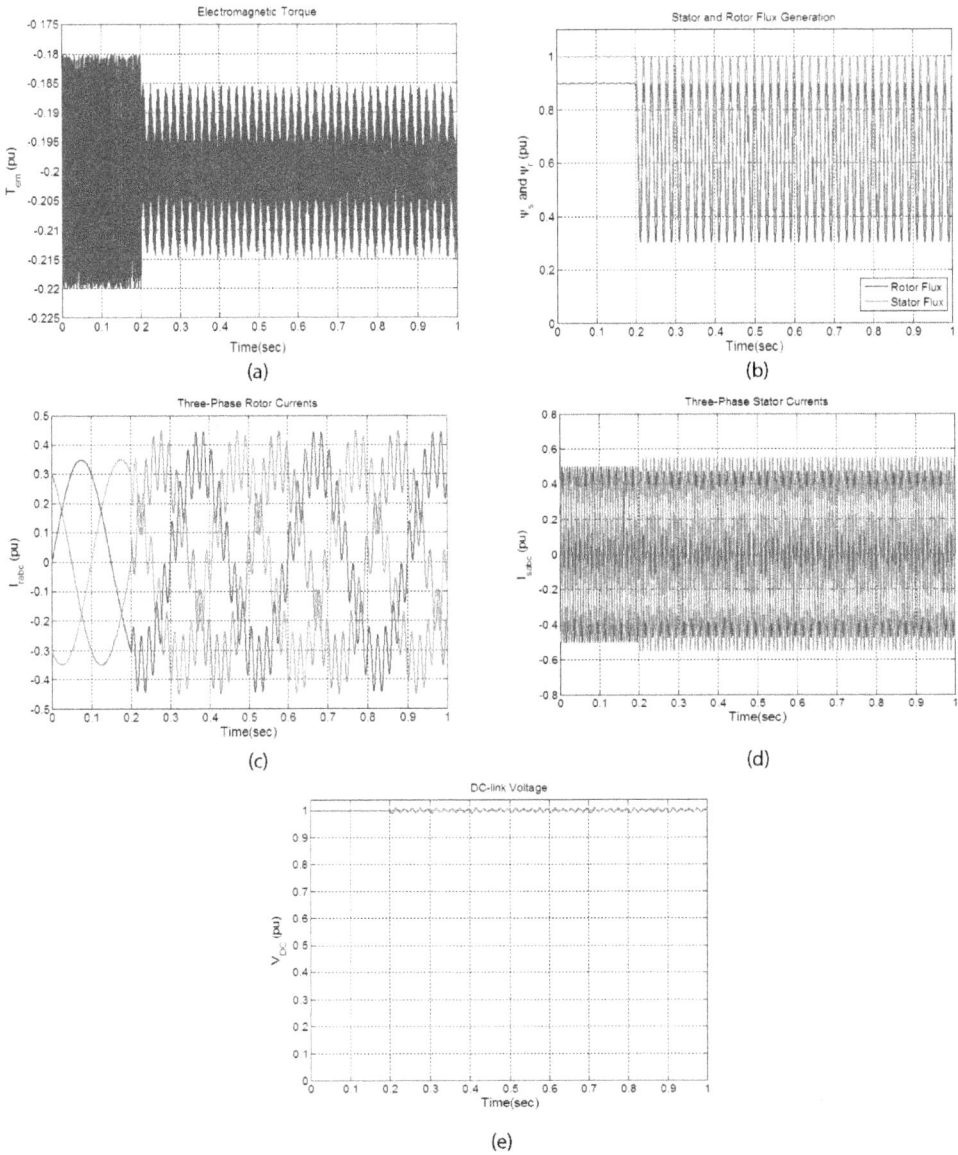

Figure 7. (a) Torque of DFIM with proposed rotor flux reference generation, (b) stator and rotor flux of DFIM with proposed rotor flux reference generation, (c) rotor currents of DFIM with proposed rotor flux reference generation, (d) stator currents of DFIM with proposed rotor flux reference generation, and (e) DC-link voltage of DFIM with proposed rotor flux reference generation.

Compared to the rotor currents generated by without flux reference generation scheme for longer voltage dip as shown in **Figure 6(d)**, the rotor currents for with reference generation scheme has lesser over currents, less severe and operate within the limits as shown in **Figure 7(c)**.

Likewise, the stator currents are also within the limits as showed in **Figure 7(d)**.

Figure 7(e) shows the DC-link voltage with oscillatory behavior due to fault and the oscillations are unbalanced and sinusoidal as mentioned previously.

5. Conclusions

In this chapter, rotor flux reference generation control strategy has been developed. Various cases have been considered such as: (a) with rotor flux reference generation without voltage dip, (b) without rotor flux reference generation with voltage dip, (c) with rotor flux reference generation with voltage dip, (d) without rotor flux reference generation during longer voltage dip, and (e) with rotor flux reference generation during longer voltage dip. Results are presented to validate the proposed scheme. From the results, it is observed that, during voltage dip, the rotor flux reference generation control scheme along with the DTC scheme eliminates the high peaks in torque with reduced stator and rotor currents, and also eliminates the necessity of crowbar during low voltage dip; the scheme makes the possibility of DFIG being connected to the grid even during fault.

Author details

Gopala Venu Madhav[1]* and Y. P. Obulesu[2]

*Address all correspondence to: venumadhav.gopala@gmail.com

1 Anurag Group of Institutions, Hyderabad, Telangana State, India

2 VIT University, Vellore, Tamil Nadu, India

References

[1] Lopez J, Gubia E, Sanchis P, Roboam X, Marroyo L. Wind turbines based on doubly fed induction generator under asymmetrical voltage dips. IEEE Transactions on Energy Conversion. 2008 Mar;**23**(1):321-330

[2] Morren J, De Haan SW. Ridethrough of wind turbines with doubly-fed induction generator during a voltage dip. IEEE Transactions on Energy Conversion. 2005 Jun;**20**(2):435-441

[3] Seman S, Niiranen J, Arkkio A. Ride-through analysis of doubly fed induction wind-power generator under unsymmetrical network disturbance. IEEE Transactions on Power Systems. 2006 Nov;**21**(4):1782-1789

[4] Slootweg JG, Kling WL. Modeling of large wind farms in power system simulations. In: Power Engineering Society Summer Meeting, 2002. IEEE, Jul 25, 2002. Vol. 1, pp. 503-508

[5] Abad G, Lopez J, Rodriguez M, Marroyo L, Iwanski G. Doubly Fed Induction Machine: Modeling and Control for Wind Energy Generation. New Jersey: IEEE Press, John Wiley & Sons; 2011 Sep 28

[6] Ekanayake JB, Holdsworth L, Wu X, Jenkins N. Dynamic modeling of doubly fed induction generator wind turbines. IEEE Transactions on Power Systems. 2003 May;18(2):803-809

[7] Holdsworth L, Wu XG, Ekanayake JB, Jenkins N. Comparison of fixed speed and doubly-fed induction wind turbines during power system disturbances. IEEE Proceedings-Generation, Transmission and Distribution. 2003 May 1;150(3):343-352

[8] Seman S, Niiranen J, Kanerva S, Arkkio A. Analysis of a 1.7 MVA doubly fed wind-power induction generator during power systems disturbances. Proceedings of NORPIE 2004. Jun 2004;14:1-6

[9] Morren J, De Haan SW. Short-circuit current of wind turbines with doubly fed induction generator. IEEE Transactions on Energy Conversion. 2007 Mar;22(1):174-180

[10] Xiang D, Ran L, Tavner PJ, Yang S. Control of a doubly fed induction generator in a wind turbine during grid fault ride-through. IEEE Transactions on Energy Conversion. 2006 Sep;21(3):652-662

[11] López J, Gubía E, Olea E, Ruiz J, Marroyo L. Ride through of wind turbines with doubly fed induction generator under symmetrical voltage dips. IEEE Transactions on Industrial Electronics. 2009 Oct;56(10):4246-4254

[12] Pannell G, Atkinson DJ, Zahawi B. Minimum-threshold crowbar for a fault-ride-through grid-code-compliant DFIG wind turbine. IEEE Transactions on Energy Conversion. 2010 Sep;25(3):750-759

www.ingramcontent.com/pod-product-compliance
Lightning Source LLC
Chambersburg PA
CBHW081235190326
41458CB00016B/5788